COMMITTEE ON REVEALING CHEMISTRY THROUGH ADVANCED CHEMICAL IMAGING

Chairperson

NANCY B. JACKSON, Sandia National Laboratories, Albuquerque, NM

Members

PIERRE CHAURAND, Vanderbilt University, Nashville, TN
JULIA E. FULGHUM, University of New Mexico, Albuquerque
RIGOBERTO HERNANDEZ, Georgia Institute of Technology, Atlanta
DANIEL A. HIGGINS, Kansas State University, Manhattan
ROBERT HWANG, Sandia National Laboratory, Albuquerque, NM
KATRIN KNEIPP, Harvard Medical School and Massachusetts General Hospital, Boston, MA
ALAN P. KORETSKY, National Institute of Neurological Disorders and Stroke, Bethesda, MD
CAROLYN A. LARABELL, Lawrence Berkeley National Laboratory, Berkeley, CA
STEPHAN J. STRANICK, National Institute of Standards and Technology, Gaithersburg, MD
WATT W. WEBB, Cornell University, Ithaca, NY
PAUL S. WEISS, Pennsylvania State University, University Park
NEAL WOODBURY, The Biodesign Institute at Arizona State University, Tempe
XIAOLIANG SUNNEY XIE, Harvard University, Cambridge, MA
EDWARD S. YEUNG, Iowa State University, Ames

National Research Council Staff

KAREN LAI, Research Associate (until December 2005)
TINA M. MASCIANGIOLI, Program Officer
ERICKA M. McGOWAN, Research Associate
CHRISTOPHER K. MURPHY, Program Officer (until July 2005)
SYBIL A. PAIGE, Administrative Associate
DAVID C. RASMUSSEN, Project Assistant
DOROTHY ZOLANDZ, Director

D1318932

Acknowledgment of Reviewers

This report has been reviewed in draft form by persons chosen for their diverse perspectives and technical expertise in accordance with procedures approved by the National Research Council's Report Review Committee. The purpose of this independent review is to provide candid and critical comments that will assist the institution in making the published report as sound as possible and to ensure that it meets institutional standards of objectivity, evidence, and responsiveness to the study charge. The review comments and draft manuscript remain confidential to protect the integrity of the deliberative process. We wish to thank the following individuals for their review of this report:

Dr. Adam Hitchcock, McMaster University, Hamilton, ON, Canada
Dr. James L. Kinsey, Rice University, Houston, TX
Dr. Anders Nilsson, Stanford University, Stanford, CA
Dr. Lukas Novotny, University of Rochester, Rochester, NY
Dr. Ralph Nuzzo, University of Illinois, Urbana
Dr. George C. Schatz, Northwestern University, Evanston, IL
Dr. James W. Serum, SciTek Ventures, West Chester, PA
Dr. Jerilyn Timlin, Sandia National Laboratory, Albuquerque, NM
Dr. Zhong Lin Wang, Georgia Institute of Technology, Atlanta
Dr. Warren S. Warren, Duke University, Durham, NC
Dr. Christine Wehlburg, MITRE Corportation, McLean, VA
Dr. Ralph Weissleder, Massachusetts General Hospital, Charlestown
Dr. Daniel P. Weitekamp, California Institute of Technology, Pasadena
Dr. Gwyn Williams, Jefferson Laboratory, Newport News, VA

Dr. Nicholas Winograd, Pennsylvania State University, University Park
Dr. Wai Tak Yip, University of Oklahoma, Norman
Dr. Ahmed H. Zewail, California Institute of Technology, Pasadena

Although the reviewers listed above have provided many constructive comments and suggestions, they did not see the final draft of the report before its release. The review was overseen by **Dr. Stephen R. Leone**, University of California, Berkeley appointed by the Division on Earth and Life Studies, who was responsible for making certain that an independent examination of this report was carried out in accordance with institutional procedures and that all review comments were carefully considered. Responsibility for the final content of this report rests entirely with the authors and the institution.

Contents

Executive Summary

Scientists have long relied on the power of imaging techniques to help them see things invisible to the naked eye and thus advance scientific knowledge. In medicine, X-ray imaging and magnetic resonance imaging (MRI) have added a level of insight beyond traditional lab tests into the workings of the human body and identification of disease at its earliest stages. Microscopy, which has been in use since the sixteenth century, is now powerful enough to detect, identify, track, and manipulate single molecules on surfaces, in solutions, and even inside living cells.

Despite these advances, today's demands on imaging have grown well beyond traditional "photographic" imaging such as medical X-ray applications. The new frontiers in microelectronics, disease detection and treatment, and chemical manufacturing demand the ability to visualize and understand molecular structures, chemical composition, and interactions in materials and reactions (see example in Box 1). In fact, in many areas, including new material development and understanding of cellular function in disease and health, the great leaps forward will depend upon the development of new and innovative imaging techniques. As a result, scientists and engineers are constantly pushing the limits of technology in pursuit of chemical imaging—the ability to visualize molecular structures and chemical composition in time and space as actual events unfold—from the smallest dimension of a biological system to the widest expanse of a distant galaxy.

At the request of the National Science Foundation, the U.S. Department of Energy, the U.S. Army, and the National Cancer Institute, the National Academies were asked to review the current state of chemical imaging technology, identify promising future developments and their applications, and suggest a research and

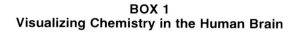

BOX 1
Visualizing Chemistry in the Human Brain

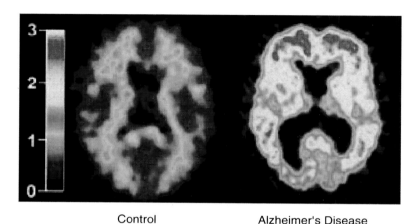

Control   Alzheimer's Disease

New positron emission tomography (PET) probes that bind specifically to amyloid plaques are promising candidates for quantifying plaque burden noninvasively. A positron-emitting chemical compound, Pittsburgh Compound B (PIB), binds specifically to amyloid plaque and can be used to image Alzheimer's disease using PET. The image on the left is the brain of a normal person, and the blue colors indicate little accumulation of PIB. The image on the right is the brain of a person with Alzheimer's disease, and the yellow to red colors indicate large accumulations of PIB and thus the presence of amyloid plaque.

SOURCE: Klunk, W.E., H. Engler, A. Nordberg, Y. Wang, G. Blomqvist, D.P. Holt, M. Bergstrom, I. Savitcheva, G.F. Huang, S. Estrada, B. Ausen, M.L. Debnath, J. Barletta, J.C. Price, J. Sandell, B.J. Lopresti, A. Wall, P. Koivisto, G. Antoni, C.A. Mathis, and B. Langstrom. 2004. Imaging brain amyloid in Alzheimer's disease with Pittsburgh Compound-B. *Ann. Neurol.* 55:306-319. Printed with permission from John Wiley & Sons, Inc.

educational agenda to enable breakthrough improvements. This report identifies the advances in chemical imaging—either new techniques or combinations of existing techniques—that could have the greatest impact on critical problems in science and technology.

A GRAND CHALLENGE FOR CHEMICAL IMAGING

A major goal for chemical imaging is both to (1) gain a fundamental understanding of complex chemical structures and processes, and (2) use that knowledge to control processes and create structures on demand. Researchers would like to be able to image a material or a process using multiple techniques across all length and time scales. Very advanced applications of chemical imaging will allow chemical images to be collected in situ (i.e., from inside the body, inside high pressure chemical reactors, or inside a cell). The ability to control complex chemical processes will require that the same techniques used for imaging can also be used to directly modulate the system under study. Reaching this grand challenge will require imaging work in areas such as self-assembly, complex biological processes, and complex materials.

For example, the self-assembly of molecules into ordered and functioning structures is a ubiquitous and spontaneous occurrence in nature: the formation of crystals; the smooth, curved surface of a drop of water on a leaf; the folding of proteins into shapes that allow them to perform specific functions. The very question of how life began, how organic molecules such as RNA or DNA first formed, may only be answered by understanding the process of self-assembly of molecules. To image the final structure of a self-assembled process, depending upon the entity formed, a scientist may use transmission electron microscopy (TEM), scanning electron microscopy (SEM), or atomic force microscopy (AFM). However, current technologies are limited. For self-assembly, opportunities to develop imaging techniques include making the methods fast enough to monitor microsecond (one millionth of a second) transformations; nimble enough to image the entire length scale from nanometers (one billionth of a meter) to millimeters; and discerning enough to image single molecules without fluorescence labeling, which impairs accessibility. In some cases, time resolution as short as femtoseconds or attoseconds will be desired.

Addressing the Grand Challenge

Understanding and controlling complex chemical processes thus requires the ability to perform multimodal imaging across all length and time scales. Complete characterization of a complex material requires information not only on the surface or bulk chemical components, but also on stereometric features such as size, distance, and homogeneity in three-dimensional space. It is frequently difficult to uniquely distinguish between alternative surface morphologies using a single analytical method and routine data acquisition and analysis. The goal of multitechnique image correlation includes extending lateral and vertical spatial characterization of chemical phases; enhancing spatial resolution by utilizing techniques with nanometer or better spatial resolution to enhance data from techniques with spatial resolutions of microns (for example, AFM or SEM combined with X-ray photoemission spectroscopy [XPS] or Fourier transform infrared spectroscopy) and facilitating correlation of different physical properties (for example, phase information in AFM with chemical information in XPS). By combining techniques that use different physical principles and record different properties of the object space, complementary and redundant information becomes available. While this gets closer to understanding and controlling complex chemical processes, further advances will also require developments in certain key areas of chemical imaging research.

AREAS OF CHEMICAL IMAGING RESEARCH

Understanding and controlling complex chemical processes also requires advances in more focused areas of imaging research. These chemical imaging techniques span a broad array of capabilities and applications. The methods are discussed in great detail within this report. Here, we briefly highlight the research and development that will best advance current capabilities—with a focus on applications in which investment would most likely lead to proportionally large returns. Succinctly summarized, the main findings of the committee are:

Nuclear Magnetic Resonance

Nuclear magnetic resonance (NMR) and magnetic resonance imaging (MRI) represent mature technologies that have widespread impact on the materials, chemical, biochemical, and medical fields. NMR and MRI are very useful tools for obtaining structural and spatial information. It is clear that in the coming years, NMR and MRI will continue to expand rapidly and continue to be key tools for chemical imaging. However, the major limiting factor for application of these techniques to a broader range of problems is their relatively low sensitivity, which is a result of the low radio-frequency energy used. Several ways that need to be explored to obtain more signal from NMR and MRI are highlighted below:

NMR Detectors

A major limiting factor of NMR and MRI is the relatively low sensitivity of their detectors. Increasing signal-to-noise ratios should be a chief focus of the efforts to improve the sensitivity of NMR and MRI detectors. In particular, this will require development of new detector insulation materials and configurations.

Hyperpolarization for NMR

Another very promising avenue for increasing sensitivity in NMR and MRI is to increase signal from the molecules being detected. A very dramatic way to do this is to couple the nuclear spins being detected by NMR to other spins with a higher polarization—such as by transferring polarization from electrons, optically pumping to increase nuclear polarization, or using parahydrogen. This is called hyperpolarization. There is a need to expand the range of techniques useful for hyperopolarizing NMR and MRI signals, as well as the range of molecules that can be hyperpolarized. Success in this area would have a great impact on all areas of NMR and MRI, from detailed structure determination to biomedical imaging.

New Contrast Agents for MRI

Contrast agents used in visualizing particular features of biological tissues such as tumors have played an important role in the development of MRI. Because MRI contrast agents are used in vivo, safety is often a major concern. Thus, improvements are needed that will improve performance of contrast agents so that they can be used in smaller quantities. This will involve developing MRI probes that have higher relaxivity, are more specific, and are deliverable to the site of action. One promising area includes MRI-active proteins or protein assemblies that are equivalent to fluorescent proteins.

Magnets for NMR and MRI Imaging Applications

The sensitivity of magnetic resonance also increases with higher magnetic fields. However, producing higher magnetic fields typically means the need for larger magnets and larger (expensive) dedicated facilities to house them. There are efforts now underway to decrease the siting requirements of high-field magnets. These efforts could decrease the size of magnets, enabling very high field NMR and MRI to transition from dedicated laboratories to widespread use for applications such as advanced oil exploration, homeland security, and environmental study. The miniaturization of high-field NMR and MRI magnets is needed to broaden the applicability of these techniques by reducing the need for dedicated facilities.

Optical Imaging

In contrast to NMR and MRI, optical spectroscopy imaging techniques utilize radiation at an energy level high enough to allow individual photons to be measured relatively easily with modern equipment at a detection sensitivity almost matched by the mammalian eye. As a result, the sensitivity and inherent temporal and spatial resolution are also increased proportionately. However, the structural information obtained from optical spectra is considerably less than that of magnetic resonance, particularly in the electronic region of the spectrum. Thus, research needs in the area of optical imaging are focused more on increasing structural information.

Optical Probes Based on Metallic Particles

In terms of the high content of chemical structural information at desired spatial and temporal resolutions, Raman spectroscopy has the potential to be a very useful technique for chemical imaging. However, a disadvantage in many applications of Raman imaging results from relatively poor signal-to-noise ratios due to the extremely small cross section of the Raman process, 12 to 14 orders of magnitude lower than fluorescence cross sections. New methodologies such as localized surface-enhanced Raman scattering (SERS) and nonlinear Raman spectroscopy can be used to overcome this shortcoming. Developing a better theoretical understanding of the radiation signals of gold and silver nanostructures, including Raman scattering, Mie scattering, and fluorescence, will enhance the applicability of optical probe microscopies. New probes composed of metal-based nanoparticles or atomic clusters need to be developed to provide improved sensitivity, specificity, and spatial localization capabilities.

Fluorescent Labels for Bioimaging

Unlike NMR and vibrational spectroscopy, electronic optical spectroscopy involves interactions with electromagnetic waves in the near-infrared, visible, and ultraviolet (UV) spectral regions. While electronic spectroscopy is less enlightening about structural information than NMR and vibrational spectroscopy, the shorter wavelengths involved allow higher spatial resolution for imaging, and its stronger signal yields superb sensitivity. Fluorescence detection, with its background-free measurement, is especially sensitive and makes single fluorescent molecules detectable. On the other hand, particularly under ambient conditions, the amount of molecular structural information that can be obtained from fluorescence imaging is limited. Organic fluorophores that bind specifically to macromolecules, metabolites, and ions provide powerful tools for chemical imaging in cells and tissues. However, the efficiencies of chemical and biological labels are hampered by photobleaching. Greater understanding of the photophysics and photochemistry of fluorescent labels, and the mechanisms of their photobleaching, need to be developed. This will broaden their applications. There is also a need to make fluorescent labels more specific, brighter, and more robust in order to probe chemical constituents and follow their biochemical reaction in cells and tissues.

Nonlinear Optical Techniques

In addition to imaging based on single-photon excited or linear Raman scattering, vibrational images can also be generated using nonlinear coherent Raman spectroscopies. The most prominent nonlinear Raman process for imaging is coherent anti-Stokes Raman scattering (CARS). Like spontaneous Raman microscopy, CARS microscopy does not rely on natural or artificial fluorescent labels, thereby avoiding issues of toxicity and artifacts associated with staining and photobleaching of fluorophores. Techniques such as CARS microscopy and other nonlinear Raman methods offer the possibility of new contrast mechanisms with chemical sensitivity, but their potential depends critically on advances in laser sources, detection schemes, and new Raman labels. Continued developments in these nonlinear approaches will enable superhigh resolution using far-field optics without the need to employ proximal probes. There is thus a need for improved ultrafast laser sources, special fluorophores, and novel contrast mechanisms based on nonlinear methods for breaking the diffraction barrier without using proximal probes.

Ultrafast Optical Detectors

Current streak camera technologies allow one to measure the lifetime and spectral features of fluorescence with subpicosecond and subnanometer resolution, but they lack the sensitivity required for single-molecule applications. Charge-coupled device (CCD) cameras, on the other hand, can provide high spectral and/or spatial resolution at high quantum efficiency, but they lack temporal resolution and near-infrared (IR) sensitivity. There is a need to develop detectors that possess the ability to measure multiple dimensions in parallel fashion, high time resolution, high sensitivity, and broad spectral range. IR and UV detector improvements, even if incremental, could catalyze new chemical imaging insights.

Electron and X-ray Microscopy

Techniques that probe samples with wavelengths much smaller than that of visible light provide high-resolution chemical and structural information below surfaces of materials. Images of atomic arrangements over a large range of length scales can be obtained using electron microscopy (EM) techniques. X-rays are able to penetrate materials more deeply than visible light or electrons and make it possible to determine the identity and local configuration of all the atoms present in a sample. Using X-rays, it is possible to image almost every conceivable sample type and gain unique insights into the deep internal molecular and atomic structure of most materials from objects as large as a shipping container to those significantly smaller than the nucleus of a single cell.

Sources for Electron Microscopy

A limiting factor in electron microscopy is the quality of the electron beam. Aberrations introduced by the optics limit both spatial resolution and analytical capabilities. There is a need to correct for the spherical and chromatic aberrations introduced by the electron optics. This will result in improved coherence of the beam and improved imaging and diffraction. In particular, these advances will permit the analysis of amorphous samples. Smaller beam sizes can also be achieved, allowing for sub-Angstrom resolution chemical analysis of samples. Development of higher-quality electron beams and short pulses of electron beams would broaden and deepen the application of electron microscopy.

Electron Microscopy Detectors

Detectors for electron microscopy are required to improve spatial and temporal sensitivity. Improved detectors will enable femtosecond time resolution and higher sensitivity and will reduce the number of electrons needed.

Optics for X-ray Microscopy

Enhanced X-ray optical systems are needed to permit imaging at higher resolution. In particular, zone plate optics, which are currently the limiting factor for scanning transmission X-ray microscopes (STXM) and full-field X-ray microscopes (TXM), need to be improved.

X-ray Detectors

The development of detectors capable of functioning on the femtosecond time scale is needed to advance X-ray imaging. Concerted efforts need to be made in developing X-ray detectors including solid-state "pixel" detectors, detectors for hard X-ray tomography through the development of scintillators that convert X rays to visible light, and detectors that image directly onto a CCD chip with column parallel readout, as well as other detector possibilities. The goal is to improve the x-ray detector's resolution, dynamic range, sensitivity, and readout speed.

Probes for X-ray Imaging

The use of X-ray microscopy to image chemical signals in biological materials requires probes that can be applied to both naturally occurring and artificially introduced molecules, particularly proteins. Current capabilities allow for only a single molecular species of protein inside a cell to be imaged. There is a need to develop the capability to simultaneously detect *multiple* proteins inside the cell through the use of multiple probes, each of which would contain a specific metal atom that could be excited at a different X-ray energy (e.g., nanocrystals containing atoms such as Ti, V, Fe, and Ni). Thus, X ray-absorbing probes that specifically detect and localize chemical signals need to be developed.

Proximal Probe Microscopies

By definition, proximal probe microscopes employ a small probe that is positioned very close to the sample of interest for the purposes of recording an image of the sample, performing spectroscopic experiments, or manipulating the sample. All such methods were originally developed primarily for the purposes of obtaining the highest possible spatial resolution in imaging experiments. Since then, many other unique advantages of these techniques have been realized. These methods are especially useful for understanding the chemistry of surfaces—for example, the electrophilicity of individual surface atoms, the organization of atoms or molecules at or near the surface, and the electronic properties of atomic or molecular assemblies. Research needs for expanding these capabilities are discussed below.

Penetration Depth

Most proximal probe imaging techniques are limited to imaging surfaces or near-surface regions. Imaging below surfaces would allow studies of chemistry at the atomic or molecular level occurring at buried interfaces and/or defects sites in the bulk of samples. There is a need for methods to be developed—for optical, X-ray, Raman, and other probe regimes—that can image at depths of a few nanometers to macroscopic distances beneath a surface for materials and life science applications. Also, there is a need for more sensitive cantilevers and stronger magnetic field gradients.

Chemical Selectivity

Many interesting materials systems are chemically heterogeneous on a wide range of length scales down to atomic dimensions. The development of chemically selective proximal probe imaging methods has played a central role in uncovering sample heterogeneity and understanding its origins. One of the best examples of the use of proximal probe methods is the chemical bonding information that has been obtained on semiconductor surfaces by scanning tunneling microscopy. However, this capability is limited in the biomedical realm and other application areas. Contrast mechanisms need to be further developed to reveal chemical identity and function in surface characterization of a wider variety of samples.

Near-field Optics

Molecular spectroscopies are restricted to length scales governed by the wavelike nature of light; specifically, spatial confinement of the source radiation is limited by diffraction to approximately one-half the wavelength of light. Near-field optical microscopy (optical proximal probe methods) overcomes this limitation and provides a means to extend optical spectroscopic techniques to the nanometer scale. However, the use of near-field microscopy to obtain chemical images of real-world samples remains hampered by issues of resolution and sensitivity. Improved probe geometries are required for high-resolution chemical imaging beyond the diffraction limit. This includes design (theory) and realization (reproducibility, robustness, mass production) of controlled geometry near-field optics.

Image Processing and Analysis

The expression "chemical image" describes a multidimensional dataset whose dimensions represent variables such as x, y, z spatial position, experimental wavelength, time, chemical species, and so forth. Image processing requires that the chemical images exist as digital images. Key aspects of imaging data collection and analysis are outlined below.

Initial Image Visualization

Frequently, the first priority for the analyst is to generate an image or images that allow for visualization of heterogeneous chemical distributions in space or time. Image visualization methods vary from simply choosing a color scale for display of a single image to methods for displaying three-dimensional datasets. For three-dimensional data, additional analysis tools are required, including the ability to extract spectra from a selected region of interest for multispectral imaging datasets or rendering a three-dimensional volume or projection for depth arrays. Analysis tools for three-dimensional visualization need to be further developed for various kinds of microscopy and surface analysis instrumentation.

Image Processing

Typically, data analysis is not considered until after an instrument is developed. This can often limit the imaging analysis or make it unnecessarily difficult. Particularly with quantitative techniques, maintaining calibration or correcting for instrument drift over time is a challenge. Integration of data analysis with instrument development will facilitate rapid acquisition and processing of images.

Multidimensional Image Processing

Present commercial multivariate analysis software is based on techniques that are more than 20 years old. It is necessary to develop better analysis and data extraction techniques for elucidating more and different kinds of information from an image. In particular, this includes user-friendly multivariate analysis tools and hyperspectral imaging deconvolution and analysis.

Integrated Real-Time Analysis

As the sophistication and multimodality of imaging instrumentation increase and as the resolution of data collections improves, it will be necessary to perform some aspects of data reduction interactively during the measurement. There is a need to develop integrated real-time analyses for automated customization of data collection, particularly in multiscale imaging applications.

Molecular Dynamics

A quantitative understanding of molecular electronic structure is vital to advances in chemical imaging. This understanding can be achieved through molecular dynamics (MD) simulations. In order to improve MD simulations, a number of specific areas need to be addressed in basic molecular dynamics theory. There is a need to develop a next generation of readily accessible, easy-to-use MD simulation packages.

All Imaging Techniques

Light Sources

Brighter, tunable ultrafast light sources would benefit many of the areas discussed in the report, particularly infrared-terahertz (between visible light and radio waves) vibrational and dynamical imaging, near-field scanning optical microscopy (NSOM), and X-ray imaging.

Miniaturization

A key route toward advancing our chemical imaging capabilities is that of miniaturization and speed of microscopic image application instrumentation.

Acquisition Speed and Efficiency

Nearly all imaging techniques would be greatly enhanced by increased data acquisition speeds. Furthermore, on-line analysis capabilities would improve the efficiency of imaging by allowing more directed investigations of samples.

Data Management

Theory must play a role in addressing the data storage and search problems associated with the increasingly large datasets generated by chemical imaging techniques.

INSTITUTIONAL CHANGE

Chemical imaging can provide detailed structural and functional information about chemistry and chemical engineering phenomena that have enormous impacts on medicine, materials, technology, and environmental sustainability. Chemical imaging is also poised to provide fundamental breakthroughs in the basic understanding of molecular structure and function. The knowledge gained through these insights offers the potential for a paradigm shift in the ability to control and manipulate matter at its most fundamental levels. A strategic, focused research and development program in chemical imaging supported by enhanced individual and multidisciplinary efforts will best enable this transformation in our understanding of and control over the natural world. This will include promoting novel approaches to funding mechanisms for chemical imaging and the development of standards for chemical image data formatting.

CONCLUSION

Imaging has a wide variety of applications that have relevance to almost every facet of our daily lives. These applications range from medical diagnosis and treatment to the study and design of material properties in novel products. To continue receiving benefits from these technologies, sustained efforts are needed to facilitate understanding and manipulation of complex chemical structures and processes. By linking technological advances in chemical imaging with a science-based approach to using these new capabilities, it is likely that fundamental breakthroughs in our understanding of basic chemical processes in biology, the environment, and human creations will be achieved.

1

Introduction

Early humans relied heavily on the ability to successfully scan their environment to hunt for food or to detect threats. While all five senses played a role in survival, vision was paramount. The human brain is designed to quickly process visual images and search for patterns, which early humans used to detect subtle changes that could indicate either hunting game or physical threats. As a result, visual images are a very "natural" and important means of obtaining information. This can be seen in everyday examples such as the preference for visual symbols over written words in traffic signs or advertising efforts to associate a specific visual image with a particular brand. In both of these examples, the choice of a symbol over words is intentional and results from the fact that the brain can grasp images faster than it can process written text.

Science has also exploited the power of imaging. A great deal of information can be obtained about a patient's condition through metabolic readings and lab tests. However, the development of medical imaging techniques such as magnetic resonance imaging (MRI) or X-ray computed tomography (CT) as standard tools for medical diagnosis has provided physicians with a new level of insight into the workings of the human body and the identification of disease at its earliest stages.

In addition, most space probes contain an imaging camera as part of their instrument package, including those operating in environments where visual images are exceedingly difficult to obtain (e.g., on the surface of Saturn's moon Titan, which is located more than 1 billion miles from the sun and is perpetually shrouded by thick clouds). These planetary images do more than provide interesting photos for public enjoyment; they also allow scientists the opportunity to make a quick determination of features of interest for further exploration, to gain insights into geological and weather modifications, and to help correlate data returned by parallel instruments on board the spacecraft.

WHAT IS CHEMICAL IMAGING?

While chemical imaging means many things to many people, concisely it is the spatial (and temporal) identification and characterization of the molecular chemical composition, structure, and dynamics of any given sample. Today's technologies and demands on imaging are growing well beyond traditional "photographic" imaging as exemplified by medical X-ray applications. To address issues such as the next generation of microelectronics technologies, disease detection and treatment, chemical manufacturing, and advanced materials development, the ability to perform spatially resolved measurements of chemical structure, function, and dynamics is vital. For example, the location and identification of atoms and molecules in the heterostructures within a state-of-the-art microprocessor are crucial to developing faster and more reliable computing architectures. Imaging the dynamic chemical processes involved in the catalytic production of chemicals is essential to improving chemical manufacturing. Imaging and tracking molecular biochemical processes is central to the development of new ways to detect and treat diseases.

Modern spectroscopic techniques rely on the interaction of light or other radiation with a sample of interest. The resulting spectra from these techniques provide vast amounts of information about molecular interactions and structures that occur in chemical processes. Even the best spectra, however, are limited in their ability to reveal the exact characteristics of a chemical reaction definitively. Most common spectroscopic methods require significant samples; for example, nuclear magnetic resonance (NMR) spectroscopy usually requires on the order of a milligram or more. While NMR is not as sensitive as many of the other techniques, in general spectra acquired using standard spectroscopic methods are the result of the accumulation of data from millions (or billions) of individual molecules. In other words, most common spectroscopic techniques provide only an approximation of what occurs between individual units in a single chemical reaction.

Chemical imaging takes advantage of a number of spectroscopic techniques (which will be discussed in depth in Chapter 3). These techniques provide the needed information about the molecular composition, structure, and dynamics of a given sample in space and time. Unlike traditional spectroscopy, however, it is now possible in certain implementations of chemical imaging to obtain images on the molecular scale, where interactions between the smallest units of structure are revealed. These advances greatly enhance the fundamental understanding of chemical interactions. As shown in Figures 1.1A and 1.1B, imaging techniques cover a wide range of time scales and penetration depths for samples of varying lateral dimension.

Development of scanning tunnel microscopy (STM) by Gerd Binnig and Heinrich Rohrer in 1981 pointed the way to breakthroughs in understanding basic chemical processes. Since then, STM and atomic force microscopy (AFM), as well as optical force microscopy proximal probes,[1] have been used to manipulate individual atoms and molecules on surfaces.

Since then, advances in instrumentation, particularly better probes, and the enormous advances in computational power promise to revolutionize chemical imaging capabilities. The potential to perform chemical imaging in real time across spatial dimensions from the nanometer to the meter scale would lead to fundamental breakthroughs in our understanding of basic chemical processes and, with this, anticipated advances in capabilities both within the chemical sciences and in a number other fields of interest.

CHEMICAL IMAGING AND FUNDAMENTAL CHALLENGES

As noted above, advances in imaging not only will benefit the chemical sciences, but also fundamental understanding in many other areas. For example, biological processes, materials, medicine, and national security provide excellent examples important application areas for advances in chemical imaging.

Biological Processes

Tremendous advances have been made in the understanding of such fundamental biological processes as cell function and protein folding. The ability to obtain in situ data on the complexity and dynamics of biological processes, however, continues to pose challenges. Because most of these processes are essentially chemical interactions, advances in chemical imaging with applications to living systems hold the potential for fundamental breakthroughs in the understanding of biological systems.

Materials

Imaging is a common technique for assessing materials. For example, airliner structural materials are often X-ray-imaged to check for hairline cracks or other signs of imminent failure. In scientific applications, advances in nanotechnology have produced a parallel need for improved methods of imaging nanomaterials on the molecular scale.

Advances in chemical imaging will have a direct impact on the ability to design, test, and alter novel materials. In addition, these advances will also contribute to the ability to control reactions at the molecular level by using various imaging modalities to project spatial and temporal information into chemical systems as well as extract information from them. This is a fundamental goal of total chemical-based synthetic processes. Exercising a greater degree of control will have a profound impact on the development of improved materials.

Medicine

Although the development of new drugs for the treatment of disease has progressed significantly beyond simple empirical "hit-or-miss" methods, comprehensive understanding of the actual interactions that take place during the delivery of medicines to a patient remains elusive. As an example, the basic principles of the physiological mechanisms by which anesthesia functions in the body (i.e., interactions between the pharmaceutical molecule and the active site within the body) are not well understood. Advances in chemical imaging, particularly in the resolution of molecule-to-molecule interactions, would help further the understanding of these processes and could contribute to innovations such as personalized medicine.

National Security

As the threat of terrorism has grown over the past decade, work toward detecting these threats has also increased. A key aspect of this work is the development of new and better sensors aimed at detecting nuclear, chemical, and biological threats. Significant hurdles complicate the ability of sensors to operate effectively in the "real world." For example, spectrometric detection is sometimes impeded by signals arising from a "noisy" background; common environmental obstacles such as smoke, moisture, or even perfume may interfere with

---

FIGURE 1.1 Imaging techniques compared by their time scales, penetration depths, and ranges of lateral dimensions. All scales are logarithmic, and all box boundaries are estimates of typical present practices and are much fuzzier than the crisp lines shown. (A) Each time span, except for electron microscopy (EM) techniques, indicates the range from the shortest time difference that can comfortably be resolved by a particular technique to the maximum duration of continuous observation. For EM techniques, the time scale indicates the estimated time required for freezing or fixing the tissue. Lateral dimensions range from the finest spacing over which separate objects can be discriminated up to the maximum size of a single field of view. (B) Depth dimensions range from the minimum thickness for an adequate signal to the maximum depth of imaging without a severe loss of sensitivity or lateral resolution. Again, lateral dimensions range from the finest spacing over which separate objects can be discriminated up to the maximum size of a single field of view. NOTE: AFM = atomic force microscopy; CARS = coherent anti-Stokes Raman scattering; FL = fluorescence microscopy at visible wavelengths; IR = infrared; MEG = magnetoencephalography; MRI = magnetic resonance imaging; NSOM = near-field optical microscopy;; PET = positron emission tomography; SERS, surface-enhanced Raman spectroscopy; STM = scanning tunneling microscopy, TIR-FM = total internal reflection fluorescence microscopy.
SOURCE: Modified version of figure supplied by Roger Tsien.

A

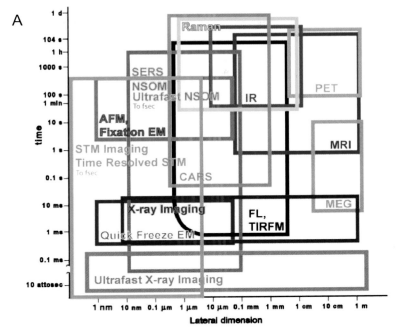

B

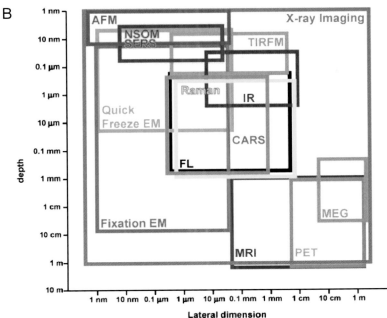

the ability to pick up a signal arising from a threat agent such as an explosive or a biological weapon. In addition, the detection of extremely dilute substances in an enormous volume of background is another difficulty that must be overcome.

While work on improved spectrometric methods continues, imaging techniques are and will continue to be powerful sensing tools to help guard against threats. Potential advantages of improved imaging techniques include the ability to assess threats in real time (allowing ample time for effective countermeasures) and the improved capability to detect threats at a distance before populations and valued assets become affected.

CHEMICAL IMAGING—WHY NOW?

A number of factors have combined over the past several years to make chemical imaging a field ripe for explosive growth. Advances in optics and nanotechnology—nanotips as probes and optical quantum dots[2] as labels[3]—have made continued improvements in imaging common. The phenomenal growth in desktop computing power—combined with the now commonplace ability to network computers—greatly diminishes the challenges once posed by storage requirements for real-time imaging. New research applications are being pursued by combining different imaging techniques to enhance imaging capabilities.

Continued advances in chemistry require more powerful techniques to visualize and manipulate matter and to efficiently manage the vast quantities of data resulting from imaging. At present, imaging techniques such as STM are limited to probing the surfaces of metals. In the short term, advances in chemical imaging that enable researchers to "see deeper" (i.e., real-time imaging below surfaces) and the ability to image "soft" materials would provide much more information than is currently available. In the long term, new capabilities in imaging will almost inevitably lead to new questions for researchers to ask, and the subsequent answers will result in the development of new capabilities as the fundamental understanding of chemical processes increases.

FOCUS OF THIS REPORT

With this wide range of applications and drivers, the approach of this report is to look for high-impact areas in which novel chemical imaging techniques can be developed either from new fundamental mechanisms of imaging or from the synergistic combination of existing techniques that will provide new information. To provide the broadest basis for these developments, an inclusive approach to chemical imaging technique development and potential has been adopted. As with all branches of science, breakthroughs will undoubtedly occur outside the scope of this report. The aim is to identify promising areas in which imaging techniques can evolve to have the greatest impact on critical problems in science and technology.

To address the current state of the art in chemical imaging and determine promising areas for advances in the field, the National Academies has undertaken the present study. The goals of the study, as described in the statement of task,[4] include:

- a review of the current state of the art in chemical imaging, including likely short-term advances;
- identification of gaps in our knowledge of the basic science that enables chemical imaging;
- identification of a grand challenge for chemical imaging;
- research required to meet this challenge;
- institutional changes that could help catalyze advances in this field.

To carry out its tasks, the Committee on Revealing Chemistry through Advanced Chemical Imaging held a series of meetings at which various imaging experts were invited to present testimonials and participate in discussions about their relevant research areas. The invited panelists included scientists and engineers from academic, government, and industrial research labs. This report is based on the information gathered at these meetings as well as the expertise of the committee members.

The committee has written its report such that these objectives are described and addressed to multiple audiences. For the nonscientist, the report seeks to describe the importance of chemical imaging not only in the chemical sciences but also for practical applications beyond the chemical research laboratory. For the student considering study in chemistry, the report aims to show that both near- and long-term advances in chemical imaging hold the potential to fundamentally alter our understanding of how a chemical reaction occurs and, perhaps more important, of what new capabilities this knowledge can enable. For chemists and chemical engineers, the report should serve as the collective judgment of experts in the field that can be used to identify new capabilities that are needed in chemical imaging and areas of research that offer the best promise of new imaging capabilities.

Chapter 2 presents a series of scientific applications of chemical imaging capabilities. Through case studies, the use of chemical imaging is detailed, including the limits on chemical imaging techniques currently in use and a discussion of the developments in chemical imaging required to fully address these scientific applications. These are discussed in the context of both short-term and long-term goals. In Chapter 3, current imaging techniques are presented in a higher level of technical detail. Furthermore, a discussion is provided of (1) the possibilities of current techniques; (2) desirable imaging tools that currently do not exist; and (3) the practical steps necessary to acquire new imaging techniques. Chapter 4 presents the committee's key findings and recommendations, which are offered as guidance for setting priorities and mapping plans toward fundamental breakthroughs in areas of imaging research as well as other areas that impact development of chemical imaging.

CONCLUSION

As work progresses to improve chemical imaging capabilities, the fundamental challenges will be to observe, understand, and control the spatial and temporal evolution of single molecules, molecular assemblies, and chemical pathways in complex, heterogeneous environments.[5] Achieving these goals will require answers to the following questions:

- How can single-molecule events be imaged in functional detail, rather than imaging the average of a collection of molecules?
- Is it possible to use chemical imaging to differentiate between intrinsic molecular behavior and cases in which molecular behavior is influenced by the environment (e.g., healthy versus diseased tissue structure or function)?
- Is it possible to control the position and/or reactivity of chemical reactions and behavior?
- What can be understood about mapping dynamics or dynamic interactions in chemical reactions?
- What role do natural processes such as self-assembly, dynamics, and environment play in controlling chemistry?
- Can chemical imaging provide insights into biological and chemical processes that inform each other?

By linking technological advances in chemical imaging with a science-based approach to using these new capabilities, it is likely that fundamental breakthroughs in our understanding of basic chemical processes in biology, the environment, and man-made creations will be achieved.

NOTES

1. The term "probe" or "proximal probe" used in this document refers to any of the wide variety of tips used in tunneling, force, and near-field optical microscopies. That is, a metallic, semiconducting, or optical-fiber probe is positioned in close proximity to a sample for the purposes of recording images.

2. Kim, S., Y.T. Lim, E.G. Soltesz, A.M. De Grand, J. Lee, A. Nakayama, J.A. Parker, T. Mihaljevic, R.G. Laurence, D.M. Dor, L.H. Cohn, M.G. Bawendi, and J.V. Frangioni. 2004. Near-infrared fluorescent type II quantum dots for sentinel lymph node mapping. *Nat. Biotechnol.* 22: 93-97.

3. In this report, the term "label" or "marker" will be used to refer to molecules or nanoparticles that covalently or otherwise chemically interact with a sample.

4. The full statement of task for this study is given in Appendix A.

5. Walter Stevens, Division of Chemical Sciences, Geosciences, and Biosciences, U.S. Department of Energy, presentation to the committee.

2

Utilizing Chemical Imaging to Address Scientific and Technical Challenges: Case Studies

This chapter provides a series of real-life "case studies" to help illustrate a grand challenge for chemical imaging. Before presenting the case studies, the grand challenge and a brief introduction to imaging techniques are discussed. More technical information about specific imaging techniques is provided in greater detail in Chapter 3.

A GRAND CHALLENGE FOR CHEMICAL IMAGING

Chemical imaging helps us to answer difficult questions, especially when these questions occur in complex chemical environments. At present, imaging lies at the heart of our high-technology industry in terms of process development and quality control. The ability to image the interior of the human body with techniques such as ultrasound and magnetic resonance imaging (MRI) has revolutionized medical diagnosis and treatment. Satellite imaging is now an indispensable tool in climate prediction and modeling. Use of remote imaging is crucial to our national security. Our capability to image will in many ways define our scientific, technological, economic, and national security future.

Clearly, advances in chemical imaging capabilities will result in more fundamental understanding of chemical processes. In this chapter, chemical imaging is addressed in the context of an overarching goal to understand and control complex chemical processes.

UNDERSTANDING AND CONTROLLING
COMPLEX CHEMICAL PROCESSES

Understanding and controlling complex chemical processes requires the ability to perform multimodal imaging across all length and time scales. That is, researchers would like the capability to image a material or a process using multiple techniques, including those that can "focus" on a particular aspect of the material or process (through varying length scales), as well as capture images at appropriate time dimensions to acquire necessary information.

An overarching objective for future breakthroughs using chemical imaging techniques is to gain a fundamental understanding and control of these complex chemical structures and processes. While this is the grand challenge for chemical imaging, more specific requirements need to be addressed in order to meet this comprehensive challenge. These include: understanding and controlling self-assembly, complex biological processes, and complex materials. Each of the challenges is amplified further below.

Understanding and Controlling Self-Assembly

The self-assembly of small molecular units into larger structures is a common and important occurrence in nature. In the biological realm, proteins and RNA fold into specific functional conformations. Cells divide and communicate with each other by rearranging subcellular units. Some theorists hypothesize that the spontaneous formation of lipid vesicles is responsible for the beginning of life. Outside biology, we marvel at the growth of snowflakes. We find numerous uses for soap and liquid-crystal displays. We make materials with varying properties by tuning the degree and the nature of aggregation. Indeed, many proposed methods for creating nanomaterials are based on self-assembly.

Molecular assemblies are formed through strong and weak chemical forces. Understanding the types, magnitudes, directions, and distances associated with these interactions is thus of fundamental and practical importance. Chemical imaging can elucidate many of these processes by providing spatial and temporal relationships among the interacting units. We would like, at one extreme, to follow the rotation, formation, and breakage of individual bonds and, at the other, to investigate cooperative effects and sequences of events over extended domains. The same or different small assemblies can be tracked as they grow into larger assemblies. In addition, chemical transformations within these structures can be monitored to elucidate environmental effects on reactivity and ultimately can be controlled by the patterned exposure to electromagnetic radiation and other fields.

To gain better understanding of and to control molecular assembly processes, one needs chemical imaging techniques that can follow interactions at a broad range of length and time scales. During assembly, it would be advantageous to record inter- and intramolecular orientations and distances at picosecond to second time scales, measure the forces between selected pairs of atoms or between selected molecular domains, and detect proximal versus long-range ordering of complexes. Once the self-assembly process is complete, imaging could be employed to follow single-molecule reactions within these structures. There are a few published examples of the monitoring of DNA synthesis and hybridization. However, major advances in imaging tools will be required to tackle the whole spectrum of molecular self-assembly processes.

Understanding and Controlling Complex Biological Processes

In the postgenomic era, there is a pressing need to functionally annotate the products of the many sequenced genes whose functions are unknown. Exploring the proteome represents a mammoth task. Although the number of genes encoded in the genomes of higher organisms has turned out to be fewer than originally thought (tens of thousands for mammals), the complexity introduced during cell development and gene expression is enormous. Combinatorial reorganization of gene fragments during immune cell development, alternate splicing pathways after transcription, and posttranslational modification of proteins and the resulting chemical heterogeneity result in millions of functionally distinct protein species. Add to this the extensive interplay of the many metabolic intermediates and connected pathways and the complexity increases even farther. This inherent complexity and heterogeneity, which is in many respects the hallmark of a living system, puts very serious limits on the utility of traditional biochemical methodologies that are based on the separation and isolation of components. A cell is much more than a list of gene products and small molecules. Just as important as the chemical formula of each component is a detailed understanding of where it is, at what time, and with what partners. Although generating a complete four-dimensional map of cellular (and ultimately organismal) complexity at the molecular level is currently beyond our capability, this is the long-range goal of chemical imaging in the realm of biology. Clearly, much has to be done to achieve this goal, but many of the fundamental concepts and tools have been or are being developed now.

From low-energy radio waves that tickle the states of nuclei, to infrared light that captures the nature and energies of chemical bonds, to visible light that probes electronic structure, to high-energy X-rays and electrons that report on electron density, spectroscopy provides detailed information and generally does so in a spatially and temporally patterned way. Scanning probe microscopy, while still largely an in vitro approach, adds an additional dimension in which mechanical and electrical probes can be applied directly. Everything from whole organisms to individual biomolecules has been imaged with these kinds of techniques. The challenge now is to come to grips with the chemical, spatial, and temporal heterogeneity involved—monitoring many molecules, molecular species, or whole cells simultaneously and thereby determining in detail the complex interactions and networks that are the chemical essence of life. Thus, there are issues of scale and a dramatic need for multiscale approaches that allow one to place the chemistry within the context of the overarching biological system.

As our ability to probe with high resolution has improved, a number of researchers have begun to consider reversing the direction of information transfer, using the same concepts and tools inherent in chemical imaging to project information into biological systems, thereby controlling their function. A somewhat crude example of this approach is laser surgery in which specific cells, or even small parts of cells, are ablated with a focused laser beam. A more sophisticated approach that has recently become possible is to specifically turn genes on or off with light, giving complete control of gene expression within a population of cells as a function of both space and time. In general, the concept of refitting our molecular imaging probes to become "full-duplex" molecules, functioning both to report on the environment that surrounds them and to manipulate that environment in an externally controlled way, is an idea that is just taking form and provides new vistas both for fundamental research in biology and for environmental, medical, and synthetic applications.

Understanding and Controlling Complex Materials

In a high-tech society, the quality of life, economic potential, and security often rest on its ability to predict and control the properties of materials. These properties can range from the common (porous, dielectric, high-strength, magnetic, chemically reactive) to the exotic (superconductivity, superlattice, superfluidity, giant magnetoresistance). In complex materials, these properties, both exotic and common, are generally determined and controlled by the degree of coupling between the components that make up the material and their resulting level of complexity (e.g., chemical and physical heterogeneity, composition, phase, morphology). Often, the degree to which we can successfully harness a particular property or phenomenon into new technologies is based largely on our knowledge and understanding of material systems at or below the size scale of the constituents and components that constitute them. For example, the discovery of new physical phenomena such as superconductivity is only the first step in what can be a long process to bring a discovery to technological relevance or commercialization. While the properties of superconductivity hold the promise of revolutionizing everything from transportation to medicine, the technological and economic impacts will go unrealized without advances in our understanding of and improvements in superconducting materials. However, progress in understanding these systems has been hampered by the absence of chemical imaging and dynamics tools that can provide nondestructive, real-time, three-dimensional imaging with relevant resolution.

In all types of materials, phenomena such as fracture, creep, segregation, roughening, and delamination ultimately determine the utility of a material for a given application. For example, by controlling the onset of fracture, a potato chip bag can be an effective, high-strength, low-porosity container that keeps chips from going stale, while at the same time allowing a child to rip open (fracture) the package with ease. Phenomena such as fracture mechanics are useful but not always well understood; as a consequence, these and other materials advances come by way of much trial and error. This is due in large part to the lack of suitable analytical instrumentation that can image over several length scales such things as the formation of stress (morphology contrast) and the resulting phenomena (fracture). While improved performance (e.g., high directional strength) is often an important driver in technology development, materials advances that make the technology affordable and more durable offer value and motivation as well.

In addition, complex materials can comprise several unique components (metals and nonmetals, liquids and solids, magnetic and nonmagnetic materials) that, when combined, generate a material whose properties are altered or totally distinct from those of the original. An example of this is the thin-film material systems that exhibit giant magnetoresistivity (GMR). Any of the thin films acting alone would exhibit no unique or exotic properties. However, when several materials are combined in a precise manner, the phenomenon of GMR is observed, and high-density data storage is realized.

Through advances in chemical imaging capability, we will increase both our basic understanding of the phenomena that determine the utility of complex materials and our ability to control or "tune" a material's properties. In this way, we will go from using the inherent properties of traditional material, (e.g., the strength of steel) to programming particular properties, such as low weight and high strength, into engineered materials that are tailored for a given application.

IMAGING TECHNIQUES

The development of multiple imaging techniques provides researchers with powerful tools to probe multiple aspects of chemical problems. A more detailed discussion of these techniques is provided in Chapter 3; however, the techniques are introduced briefly here.

Optical Techniques and Magnetic Resonance

Techniques employing the ultraviolet (UV), visible, and near-infrared parts of the spectrum have the advantage of high sensitivity (single photon), high time resolution (femtoseconds), and moderate spatial resolution (on the order of 100 nm). Structural information is obtainable by infrared to radio-frequency techniques (e.g., magnetic resonance). Together, these techniques have enabled the visualization of individual molecules and the measurement of excited state dynamics from such molecules on the picosecond time scale. It is also possible to follow the time course of chemical reactions on the femtosecond time scale when whole populations can be synchronized by light. Confocal detection and non-linear excitation have made it possible to follow the dynamics of complex chemical systems (such as cells and tissues) using multiple probes and in three dimensions. As a whole, these technologies have also made it possible to optically pattern chemical reactivity with very high spatial resolution in three dimensions. Imaging well below the surface of an object (e.g., deep tissue imaging) remains a challenge in optical spectroscopy, but could be substantially improved with the production of labels absorbing and/or emitting farther to the red.

Vibrational imaging using Raman scattering and infrared (IR) absorption provides something like a structural "fingerprint" of matter as it is determined by the kinds of atoms, their bond strengths, and their arrangements in a specific molecule. Recent developments based on a combination of modern laser spectroscopy, scanning probe techniques, and nanotechnology provide capabilities for sensitive vibrational imaging at the single-molecule level. These developments also provide capabilities at nanoscale lateral resolution, where linear and nonlinear Raman scattering is exploited in enhanced and strongly confined local optical fields of tailored nanostructures.

Electron Microscopy, X-rays, Ions, and Neutrons

With wavelengths that are about 1,000 times smaller than that of visible light, electrons provide a high-resolution probe of chemical and structural information below surfaces of materials. Images of atomic arrangements over a large range of length scales can be obtained using electron microscopy (EM) techniques. Although significant limitations to their use exist (e.g., the need for a vacuum to produce and transmit electrons, electron beam damage to samples), EM techniques have had a tremendous impact on fields ranging from condensed matter physics to structural biology.

X-rays are able to penetrate materials more deeply than visible light or electrons and make it possible to determine the identity and local configuration of all the atoms present in a sample. Using X-rays, it is possible to image almost every conceivable sample type and gain unique insights into the deep internal molecular and atomic structure of most materials from objects as large as a shipping container to those significantly smaller than the nucleus of a single cell.

Proximal Probes (Force Microscopy, Near Field, Field Enhancement)

Proximal probe microscopes employ a variety of materials such as tungsten wire (scanning tunneling microscopy), silicon nitride pyramid and cantilever (atomic force microscopy), or optical fiber (near-field optical microscopy) in close proximity to the sample of interest for the purposes of recording an image of the sample, performing spectroscopic experiments, or manipulating the sample. All such methods were originally developed primarily for the purpose of obtaining the highest possible spatial resolution in imaging experiments. Since then, many other unique advantages of these techniques have been realized. These methods are especially useful for understanding the chemistry of surfaces—for example, the electrophilicity of individual surface atoms, the organization of atoms or molecules at or near the surface, and the electronic properties of atomic or molecular assemblies.

Processing, Analysis, and Computation

Processing, analysis, and computation are not imaging techniques per se but, rather, play a fundamental role in enhancing their capabilities. In addition, computational methods, particularly when applied to computer modeling and simulation, extend imaging capabilities to address problems that have not or cannot be addressed using standing imaging techniques.

CASE STUDIES

A series of real-life "case studies" is presented to illustrate the importance of the technical issues that have been introduced in this chapter and show how chemical imaging can contribute to understanding them. The examples are not meant to serve as an exhaustive list of all problems that can be addressed with advances in chemical imaging. Instead, they have been included to focus on current capabilities and limitations and offer insights into where breakthroughs are needed to increase the capabilities and potential for chemical imaging. More technical information about specific imaging techniques is provided in greater detail in Chapter 3.

Case Study 1: Mobile Crystalline Material-41 (MCM-41)

MCM-41 is an interesting self-assembled material[1] that has a wide range of applications. The starting material is a monomeric surfactant, cetyltrimethylammonium bromide (CTAB), similar to those used as detergents. With a long hydrophobic end and a hydrophilic head, the monomers form spherical micelles that have a hydrophobic core and a hydrophilic surface when the concentration of the monomer surpasses the critical micellar concentration. At higher concentrations, the micelles rearrange into cylindrical rods. At still higher concentrations, the rods self-assemble into hexagonal arrays. After the introduction of silicate molecules, the arrays form silica particles that possess well-defined shapes. By adding agents that alter the hydrophobicity or hydrophilicity of various parts of the structures, one can control each step of the self-assembly process. The result is a mesoporous structure with tunable pore size, variable channel length, and predictable shape (Figure 2.1 and Figure 2.2).[2]

MCM-41 has been employed as an industrial catalyst for many years. The assembled structure is pyrolized to become a permanent inorganic matrix.

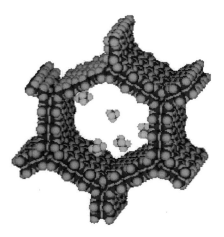

FIGURE 2.1 MCM-41 units are formed from self-assembly to create honeycomb structures that can be functionalized on the inside (light blue) to create confined catalytic sites. SOURCE: Courtesy of Victor S. Lin, Iowa State University.

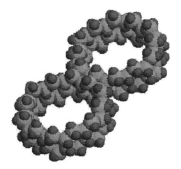

FIGURE 2.2 These honeycomb units further assemble into larger structures that can include worms, spheres, ovals, and so on, depending on preparative conditions. SOURCE: Courtesy of Victor S. Lin, Iowa State University.

Functionalization of the matrix allows incorporation of a variety of catalytic activities into the material. Recently, procedures were developed to add functional groups that are electrostatically or hydrophobically attractive to the ammonium surfactant head groups and are able to compete with silicate anions during self-assembly. This has led to a class of mesoporous materials that are functionalized only on the inside of the pores. Highly selective polymerization and cooperative catalytic systems have been developed from these materials.[3] Furthermore, by incorporating caps onto the pores, chemical reagents can be stored in the channels, to be released simply by detaching the caps at the desired time and location.[4] This scheme holds promise as a controlled drug and gene delivery protocol.

Chemical Imaging Technique(s) Involved

Current methods used to image MCM-41 include (1) analytical transmission electron microscopy (TEM) to determine structure, size, morphology, and local chemical composition; (2) energy-dispersive X-ray spectroscopy (EDXS) in a scanning electron microscope (SEM) to determine chemical composition;[5] and (3) electron energy loss spectroscopy (EELS) for elemental analysis.[6]

Insights Obtained Using Chemical Imaging

The spatial and temporal progression of individual events involved in the formation of each type of structure can be monitored directly. A combination of imaging modes can be applied, each elucidating the process at a different length scale. Millimeter-scale variations can then be explained by nanometer-scale fluctuations. After the structures are built, single-molecule imaging can be employed to study catalytic reactions inside the nanopores.

Imaging Limitations

Limitations include the following:

- The imaging rate of current technologies is not fast enough for continuous monitoring of microsecond transformations.
- Single-molecule imaging techniques are not yet capable of monitoring several different chemical species simultaneously.
- There is a lack of technologies for imaging the length scale between optical microscopy (diffraction limit) and proximal probes.
- The need for chemical derivatization for fluorescence imaging often limits accessibility.

Opportunities for Imaging Development

Opportunities to develop imaging techniques for this application would include the following:

- Optical imaging at microsecond to nanosecond time scales per consecutive image
- One instrument for imaging the entire length scale from nanometers to millimeters
- Single-molecule imaging without fluorescence labeling

Case Study 2: Organic Electronics

Organic materials are now being employed as the active components in electronic circuitry. Perhaps the best examples of such materials are the semiconducting polymers used in polymer-based light-emitting diodes (polymer-LEDs). As a result of the successful development of extremely pure polymeric materials, polymer-LEDs are now being incorporated into commercially available display devices. Emerging applications of small-molecule, oligomeric, and polymeric organic semiconductors include their use in photovoltaics (solar cells) and organic field effect transistors. The primary benefits of such materials include the ability to manufacture moldable, flexible materials for use in large-area devices. Importantly, such materials can easily be cast as thin films, offering the potential for significant cost reductions in comparison to traditional inorganic devices.

Microscopic imaging experiments have played a key role in the development of these organic material devices and have provided detailed information on the local chemical and physical properties. They have helped researchers better understand intermolecular interactions, molecular organization within nanometer scale (and larger) domains, electronic coupling between individual molecules in the aggregate, and the mechanisms of electrical charge generation, injection, transport, and recombination. Microscopic methods will continue to provide vital information on molecular- to micrometer-length scales for both existing and emerging materials. Examples are shown below (Figures 2.3-2.5, respectively): a

FIGURE 2.3 *Left*: urea-substituted thiophenes on a graphite surface. *Right*: a model.
SOURCE: Reprinted with permission from Gesquiere, A., M.M.S. Abdel-Mottaleb, S. De Feyter, F.C. De Schryver, F. Schoonbeek, J. van Esch, R.M. Kellogg, B.L. Feringa, A. Calderone, R. Lazzaroni, and J.L. Bredas. 2000. Molecular organization of bis-urea substituted thiophene derivatives at the liquid/solid interface studied by scanning tunneling microscopy. *Langmuir* 16:10385-10391. Copyright 2000 American Chemical Society.

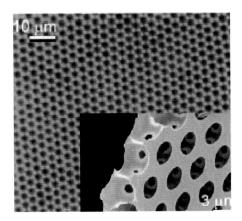

FIGURE 2.4 Honeycomb structure formed from block copolymer films.
SOURCE: Reprinted from de Boer, B., U. Stalmach, P. F. van Hutten, C. Melzer, V.V. Krasnikov, and G. Hadziioannou. 2001. Supramolecular self-assembly and opto-electronic properties of semiconducting block copolymers. *Polymer* 42: 9097-9109. Copyright 2001 with permission from Elsevier.

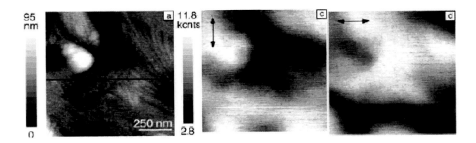

FIGURE 2.5 NSOM topography and polarized luminescence from poly(dihexylfluorene) (an organic semiconductor) film.
SOURCE: Reprinted with permission from Teetsov, J.A. and D.A. Vanden Bout. 2001. Imaging molecular and nanoscale order in conjugated polymer thin films with near-field scanning optical microscopy. *J. Am. Chem. Soc.* 123:3605-3606. Copyright 2001 American Chemical Society.

scanning tunneling microscopy (STM) image of and model for organized thiophene monolayers deposited on a graphite surface; an image of organized honeycomb structures formed in a film prepared from block copolymers of poly(phenylene vinylene)-poly(styrene) showing micrometer-scale phase separation of the component polymers as seen by fluorescence microscopy and SEM (inset); and near-field scanning optical microscopy (NSOM) topography and polarized fluorescence excitation images of annealed poly(fluorene) films showing aggregated fibrous film structures.

Chemical Imaging Technique(s) Involved

Imaging of organic electronics employs conventional fluorescence and confocal microscopies, single-molecule spectroscopy,[7] scanning and transmission electron microscopies,[8] and several different proximal probe techniques.[9,10] Optical microscopies provide direct information on spatial variations in the spectroscopic properties of the materials (i.e., due to aggregation), along with evidence for variations in the chemical composition and valuable data on their photochemical reactivities and molecular photophysics. Electron microscopy provides valuable information on nanometer and larger structures patterned within their films. Force microscopy and STM provide valuable data on organization and electronic structure on angstrom-to nanometer-scale distances, while near-field optics provides high resolution spectroscopic data with sub 50-nm spatial resolution and subnanosecond time resolution.

Insights Obtained Using Chemical Imaging

As shown in the above figures, chemical imaging has provided detailed information on molecular (self)-organization in these materials, as well as on overall film morphology and the quality of structures templated or lithographically prepared in their films. Chemical imaging methods have also provided detailed information on the optical properties of these materials, allowing a deeper understanding of the influences of inter- and intramolecular electronic coupling,[11,12] and on charge carrier dynamics and trapping.[13,14,15]

Imaging Limitations

Direct chemical information with resolution on molecular length scales cannot yet be obtained on functioning devices or even on samples closely approximating functional materials. The vast majority of high-spatial-resolution images that have been recorded have been obtained on specially prepared samples consisting of single molecular layers on well-ordered substrates. The covering electrodes and ancillary films used in functional devices also routinely present a problem in the imaging of such materials because high-resolution proximal probes cannot then be used to image their internal surfaces directly. Although conventional optical imaging methods can "see below the surface" in such samples, the resolution of these techniques is insufficient to access direct molecular information.

Opportunities for Imaging Development

Development of new imaging methods that can probe the properties of functional and even functioning[16] devices with high (molecular-scale) resolution is necessary to fully understand the organizational properties of these materials, how molecular organization influences device performance, and the detailed chemistry by which such devices fail over time. An important challenge here involves the development of methods that can image beneath the electrodes between which the materials are sandwiched, with depth discrimination capabilities. A further challenge involves implementation of techniques that provide clear chemical information (i.e., Raman, IR, or other vibrational imaging techniques) in these same imaging modalities, without sacrificing spatial resolution. Finally, for the preparation of ultimate device structures, advanced lithographic procedures based on some of these same microscopic methods will be required for the controlled fabrication of molecular architectures with optimal optical and electronic properties.

Case Study 3: Imaging Alzheimer's Disease: Chemical and Molecular Imaging of the Brain from Molecules to Mind

In the past ten years, we have made significant progress in our understanding of how the brain functions both in health and in disease. Much of this progress comes from discoveries of how to apply powerful existing technologies toward imaging the brain. Mass spectrometry and electron microscopy, for example, have been employed on brain tissue sections for the respective purposes of mapping on the micrometer scale both inorganic and organic compounds such as calcium, phospholipids, and proteins. These methods have also been used to image molecular events in structures as small as synapses. Novel imaging techniques have been developed for use on live specimens. In neurons, for example, it is possible to simultaneously image several cellular processes, such as calcium- and zinc-signaling pathways, through the use of multiphoton fluorescence spectroscopy. Another example is the now-routine clinical use of positron emission tomography (PET) and MRI spectrometers for low-resolution brain imaging on the millimeter scale. These instruments aid the diagnosis of lesions induced by strokes or the precise localization of gliomas. The ability to map areas of the brain that are active during complex tasks with functional MRI is revolutionizing cognitive psychology. Indeed, a whole field has developed within radiology called molecular imaging. Molecular imaging is the result of using traditional radiological imaging tools combined with the knowledge of specific biological processes gleaned from molecular biology to increase the range of information available from imaging. Results from chemical imaging are a key component in the development of molecular imaging.

Almost every problem faced in trying to understanding the normal functioning of the brain and pathophysiological processes in the brain requires the following steps. First, information must be acquired about the detailed structure and composition of key molecules in the brain. Second, by using this information, the dynamics of cell structure and function must be inferred. Finally, these inferences must then be integrated into an understanding of the complex functioning of the brain. Nowhere are the challenges greater than in trying to understand neurodegenerative diseases such as Alzheimer's disease. Alzheimer's is rapidly growing into a major public health problem in developed countries as their populations age. Early detection is critical, and currently, confirmation of the diagnosis relies on autopsy to detect the amyloid plaques that are the telltale sign of the disease. The detailed formation of amyloid plaques is critical to understanding and treating the disease. Finally, the detailed cellular pathology and ensuing effects on brain function are only beginning to be delineated. Thus, chemical imaging is critical to understanding Alzheimer's due to the need to determine molecular structure, cell structure, and communication and to integrate these into obtaining information nondestructively from the human brain.

Advances in chemical imaging techniques are enabling new information to be obtained about Alzheimer's across the full range of distance scales required. Detailed three-dimensional structures of amorphous solids that defy crystallization, such as amyloid plaques, are particularly challenging to characterize. Models for the structure have recently been deduced from solid-state nuclear magnetic resonance (NMR) studies (Figure 2.6). Detailed studies of the formation of amyloid plaques and their effect on specific neuronal structures nearby can be accomplished with multiphoton fluorescence imaging tools (Figure 2.7). New PET probes that bind specifically to amyloid plaques are promising candidates for quantifying plaque burden noninvasively (Figure 2.8). Finally, MRI can be used to show anatomical changes in the areas of the brain most affected (such as the hippocampus) as well as highlight the changes in brain function that occur in these areas during the disease (Figure 2.9).

Chemical Imaging Technique(s) Involved

A number of chemical imaging techniques are being utilized to understand Alzheimer's disease. Nondestructive imaging techniques such as MRI and PET are at the early stages of measuring amyloid plaque and changes in brain anatomy and function that occur with disease progression. Optical imaging is important to study nondestructively the development of the disease and changes in cellular structure that occur in animal models. The full range of in vitro chemical imaging techniques (e.g., electron microscopy, mass spectrometry imaging) has been used to chemically and structurally characterize the disease at high resolution. Finally, magnetic resonance methods and X-ray crystallography for structure determina-

(a)

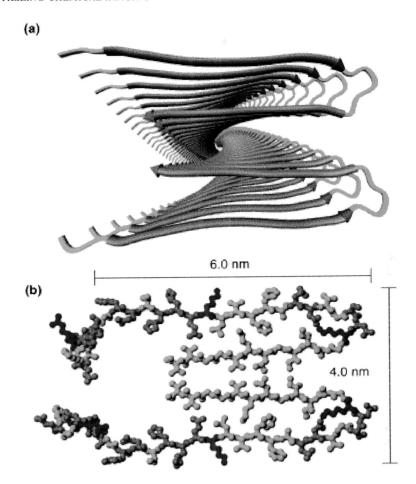

FIGURE 2.6 Solid-state NMR of the molecular structure of amyloid plaque. Model of the minimal structural unit of amyloid fibrils based primarily on solid-state NMR data. (a) Ribbon diagram of the parallel beta-sheet structure. (b) Atomic representation of the structure with colors representing side chain type (green, hydrophobic; magenta, polar; red, negatively charged; blue positively charged).
SOURCE: Reprinted from Tycko, R. 2004. Progress towards a molecular-level structural understanding of amyloid fibrils. *Curr. Opin. Struct. Biol.* 14:96-103. Copyright 2004, with permission from Elsevier.

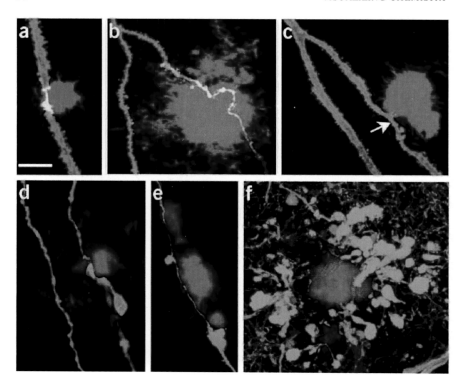

FIGURE 2.7 Two-photon fluorescence microscopy of amyloid plaque (red) and surrounding neurons (green) in the brain of a mouse model of Alzheimer's. Numerous neuronal abnormalities, including swelling and decreased densities of spines (arrow in c), could be detected.
SOURCE: Tsai, J., J. Grutzendle, K. Duff, and W.B. Gan. 2004. Fibrillar amyloid deposition leads to local synaptic abnormalities and breakage of neuronal branches. *Nat. Neurosci.* 7:1181-1183.

tion are required to understand the structure of precursors and the structure of amorphous, amyloid plaque.

Insights Obtained Using Chemical Imaging

A large amount is now known about the location of plaque formation in the brain, the pathophysiology of Alzheimer's disease, the molecular and cellular basis of the disease, and the genetic basis of the disease. All of these developments have relied on the use of chemical imaging techniques.

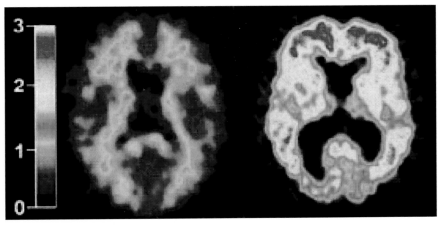

Control   Alzheimerís Disease

FIGURE 2.8 PET measurement of amyloid plaque in the human brain. A positron-emitting compound, Pittsburgh Compound B (PIB), binds specifically to amyloid plaque and can be used to image Alzheimer's disease using PET. The image on the left is the brain of a normal person, and the blue colors indicate little accumulation of PIB. The image on the right is the brain of a person with Alzheimer's disease, and the yellow to red colors indicate large accumulations of PIB and thus the presence of amyloid plaque.
SOURCE: Klunk, W.E., H. Engler, A. Nordberg, Y. Wang, G. Blomqvist, D.P. Holt, M. Bergstrom, I. Savitcheva, G.F. Huang, S. Estrada, B. Ausen, M.L. Debnath, J. Barletta, J.C. Price, J. Sandell, B.J. Lopresti, A. Wall, P. Koivisto, G. Antoni, C.A. Mathis, and B. Langstrom. 2004. Imaging brain amyloid in Alzheimer's disease with Pittsburgh Compound-B. *Ann. Neurol.* 55:306-319. Printed with permission from John Wiley & Sons, Inc.

Imaging Limitations

Despite all of the advances, we are still not able to give a definitive diagnosis or prognosis to individuals afflicted with Alzheimer's disease. In addition, it is a major hurdle to test a large number of potential drugs that may be effective in slowing the progression. Two major challenges for chemical imaging to make more rapid progress are:

1. Development of approaches that give cellular-level resolution, nondestructively, in the human brain.
2. Development of noninvasive strategies that give chemical and structural information at the same level of detail and quality as that obtained from taking samples from living tissue.

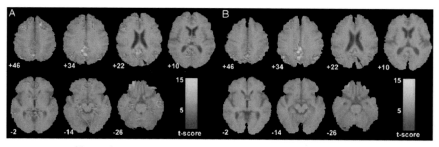

Normal Alzheimer's

FIGURE 2.9 Functional MRI of resting or default-mode brain activity in normal and Alzheimer's patients. Functional MRI detects fluctuations in brain activity at rest. A network of brain regions is activated in normal elderly people (*left*, A) as indicated by the orange-yellow regions overlaid on the MRI. This network is called the default-mode network and is altered in people with Alzheimer's (*right*, B). In particular, activity in the hippocampus and entorhinal cortex is decreased (green arrows) in Alzheimer's.
SOURCE: Greicius, M., G. Srivastava, A. Reiss, and V. Menon. 2004. Default-mode network activity distinguishes Alzheimer's disease from healthy aging: Evidence from functional MRI. *Proc. Natl. Acad. Sci. U.S.A.* 101:4637-4642. Copyright 2004 National Academy of Sciences, U.S.A.

Opportunities for Imaging Development

Major progress can be achieved in the application of chemical imaging to Alzheimer's disease and a wide range of other diseases with basic development of the imaging modalities. In particular, emphasis on increasing sensitivity and resolution, especially of noninvasive imaging modalities such as MRI, will enable images of the brain at resolutions comparable to histology. Along with the development of basic imaging technologies, it is critical to develop new imaging agents that will give greater chemical, molecular, and cellular specificity to imaging techniques. In particular, the development of optical imaging probes for detailed studies of animal models and new MRI and PET agents for human studies will greatly expand the capabilities of chemical imaging for the human brain.

Case Study 4: Nonsense Suppression Techniques for Unnatural Amino Acids in Genetic Encoding

Since experimental biology often requires that many proteins be tagged in parallel during the course of a single assay or experiment, it benefits from conjugation strategies that are simple, reliable, and easily applied to many distinguishable proteins in parallel. Nonsense suppression[17] techniques for the incorporation of unnatural amino acids into proteins (Figure 2.10) are important here because they combine the specificity of placement conferred by genetic encodability with the flexibility of multiple organic and inorganic markers. At present, these tools are well developed in bacteria and yeast, but they are at the proof-of-principle stage in higher eukaryotic cells. Support for efforts to develop these tools for use in mammalian cells would be a prudent investment and would allow the more efficient conversion of newly discovered enzymes into functional reporters that can be imaged at high resolution by fluorescence microscopy.

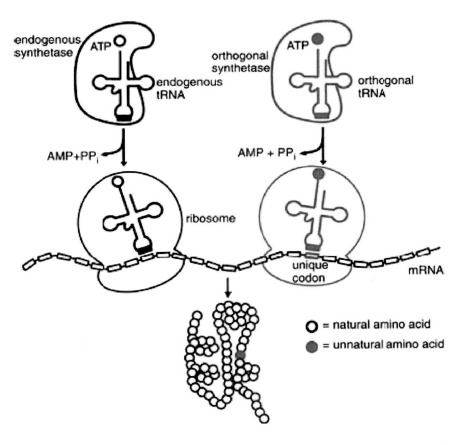

FIGURE 2.10 A general approach for site-specific incorporation of unnatural amino acids into proteins in vivo. NOTE: AMP = adenosine 5′ monophosphates; ATP = adenosine 5′-triphosphate; PP$_i$ = pyrophosphate.
SOURCE: Wang, L., and P.G. Schultz. 2005. Expanding the genetic code. *Angew. Chem. Int. Ed.* 44:34-66.

Chemical Imaging Technique(s) Involved

Bright fluorescent protein markers and accurate optical imaging provide nanoscopic spatial resolution over many orders of temporal magnitudes ranging from microseconds to several minutes.

Insights Obtained Using Chemical Imaging

It is possible to understand the biochemical dynamics of life processes by detecting and tracking individual macromolecules in living cell membranes and tissues.

Imaging Limitations

The limiting difficulty in utilizing the powerful approach to protein labeling for multiphoton chemical imaging is that higher cellular systems protect themselves by mechanisms of genetic nonsense suppression. To inhibit this barrier, it is necessary to devise means to inhibit natural nonsense suppression. Excellent but limited progress made in major laboratories[18,19] has demonstrated the power of these methods for chemical imaging of dynamic molecular processes in living systems in cells. However, broadening the applicability of these significant improvements calls for focused research efforts to reach broad applicability.

Opportunities for Imaging Development

High-resolution chemical imaging methods utilizing these diverse fluorescent markers would strongly enhance capability in analyzing molecular patterns, mobility, and interactions important in biological and materials science research.

Case Study 5: Imaging Nanochannels in Microfluidic Devices[20]

Devices based on biomolecular separation in nanochannels are expected to accelerate drug discovery, rapid diagnosis and treatment of disease, and development of vaccines. Integrated microfluidic devices are currently used to automate the generation and analysis of chemical compounds. Chemical analyses on microfluidic devices can be highly automated and can reduce the consumption of reagents by several orders of magnitude. In microchannels, the electroosmotic flow can be controlled using field effects and surface modification, but direct electrostatic manipulation of ions across the microchannels is not possible. Shrinking the dimensions of the channels down to nanometers allows direct ionic or molecular manipulation using surface charges or field effects, because the channel width then approaches the molecular diameter.

Charged solutes in electrolyte solutions that are electrokinetically driven through channels with nanoscale widths exhibit unique transport characteristics that may enable rapid and efficient separations under a variety of physiological and environmental conditions. Many biomolecules, including DNA, proteins, and peptides, are charged or can be complexed with charged surfactant molecules. Manipulating the velocity of biomolecules by variation in flow pressure or electric fields in channels of nanoscopic widths will enable efficient separations that are not possible in micro- or macroscopic channels.

Silicon-based T-chips integrate an array of parallel nanochannels with microchannels and macroscopic injection ports (Figure 2.11A). These T-chips allow the electrokinetic transport of fluorescent dyes in nanochannels to be characterized. The width of the nanochannels ranges from 35 to 200 nm, while the depth is sufficient to allow significant molecular throughput. A cross-sectional scanning electron micrograph of a small number of nanochannels in one of the T-chips is shown in Figure 2.11B. In this chip, the channels are approximately 50 nm wide by 500 nm deep and are on a 400 nm pitch. The channels are etched into a silicon wafer, which is then oxidized to present a silicon dioxide (SiO_2) surface to the fluid.

Transport of molecules through the nanochannels is studied by using confocal microscopy to monitor fluorescence from the dye molecules in the fluid. Figure 2.12 shows laser-induced fluorescence micrographs that demonstrate the difference in transport of two dyes in channels with ~50 nm (left) and ~200 nm

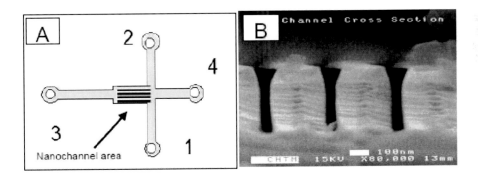

FIGURE 2.11 (A) Top view (schematic) of the integrated chips. (B) SEM image of the cross section of the nanochannel array (50 nm wide nanochannels) in a chip.
SOURCE: Courtesy of the Cancer Research Microscopy Facility, University of New Mexico Hospital; the W.M. Keck foundation; and the following individuals: Anthony L. Garcia, Linnea K. Ista, Dimiter N. Petsev, Michael J. O'Brien, Paul Bisong, Andrea A. Mammoli, Steven R.J. Brueck, and Gabriel P. Lopez.

50 nm   **200 nm**

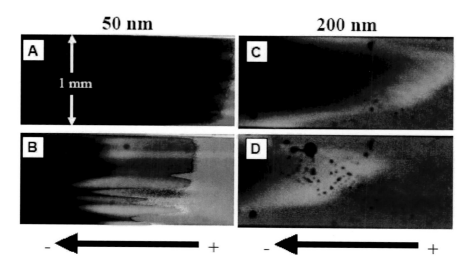

FIGURE 2.12 Sample two-color fluorescence micrographs (green = Alexa 488, red = rhodamine B) showing separation of dyes in nanochannel arrays containing channels ~50 nm wide at (A) time $t = 0$ and (B) $t = 30$ seconds, and ~200 nm wide channels at (C) time $t = 0$ and (D) $t = 25.2$ seconds.
SOURCE: Courtesy of the Cancer Research Microscopy Facility, University of New Mexico Hospital; the W.M. Keck foundation; and the following individuals: Anthony L. Garcia, Linnea K. Ista, Dimiter N. Petsev, Michael J. O'Brien, Paul Bisong, Andrea A. Mammoli, Steven R.J. Brueck, and Gabriel P. Lopez.

(right) widths. Separation of the dye fronts occurs very close to the entry to the nanochannels (in all cases less than 1 mm from the entrance). Counter to fluid flow observed in micro- and macrochannels, the negatively charged dye (Alexa 488, green fluorescence) moves more quickly toward the negatively biased electrode than the neutral dye (rhodamine B, red fluorescence).

Decreasing the nanochannel width thus leads to qualitatively new and counterintuitive behavior that can be exploited for molecular separations. Because details of the flow profiles in individual nanochannels are below the resolution limit of optical microscopy, only the average velocities of dye fronts can be monitored. Significant improvements in the lateral resolution of analytical imaging methods are required to study the transport of molecules in an individual channel.

Chemical Imaging Technique Involved

Confocal microscopy is the main imaging technique used in this research application.

Insights Obtained Using Chemical Imaging

Chemical imaging allows the behavior of molecules in nanochannels to be investigated by following the flow of fluorescent dyes. Behavior contrary to that observed in micro- and macrochannels is observed, demonstrating that this is a new and promising flow regime.

Imaging Limitations

The signal must be averaged over hundreds of nanochannels.

Opportunities for Imaging Development

Techniques that can image molecular flow in single nanochannels will enhance understanding of flow in this new regime, accelerating device development.

Case Study 6: Biological Imaging Involving Multiple Length Scales

To observe the chemistry of the human body down to the molecular details of individual cells (Figure 2.13), imaging instruments with significantly better resolving power are needed. Approximately 300 years after Anton van Leeuwenhoek built the first light microscope, this instrument continues to be the workhorse of modern biologists. By combining the developments of modern nanotechnology, optics, and computer science, the basic light microscope has been transformed and now manifests as several advanced imaging systems, such as the confocal and multiphoton microscopes. These microscopes allow observation of individual brain cells (neurons) and their axons, many of which extend from the brain to distant regions of the body.

Tagging individual proteins with fluorescent molecules allows them to be monitored in live cells, enabling rapid discovery of many intricate details about the cellular chemistry. For example, fluorescent tags on molecules that have been packaged into small vesicles have been monitored as they travel along micro-tubules within the axons via complex protein machines.

Arrays of microtubules, which are long polymers of the protein tubulin, are observed using the recently developed X-ray microscope. This new imaging technology is in its infancy but is rapidly gaining importance because of its ability to produce three-dimensional computed axial tomography (CAT) scans of single cells. With the continued development of optics, a threefold increase in resolution will soon be possible. Electron microscopes, invented in the 1930s, can also yield chemical information over a range of length scales. By examining metallic replicas of fractured cells, the structural organization of microtubules within axons, and the distinct cross-bridging proteins between adjacent polymers can be visualized. The electron microscope used in the diffraction mode reveals the molecular

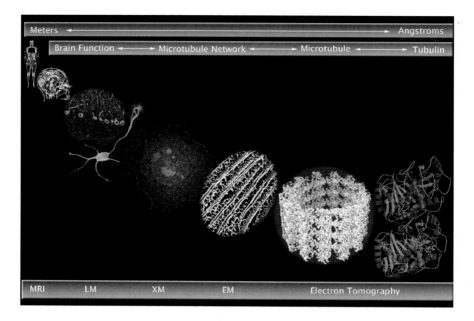

FIGURE 2.13 Instrumentation for capturing images at multiple length scales: magnetic resonance imaging (MRI), light microscopy (LM), X-ray microscopy (XM), electron microscopy (EM), and electron tomography.
SOURCE: Courtesy of Carolyn Larabell, Lawrence Berkeley National Laboratory assembled from various sources: MRI, LM, XM, microtubule network, Meyer-Ilse, W., D. Hamamoto, A. Nair, S.A. Lelièvre, G. Denbeaux, L. Johnson, A.L. Pearson, D. Yager, M.A. Legros, and C.A. Larabell, 2001. High resolution protein localization using soft X-ray microscopy. *J. Microsc.* 201:395-403; electron tomography (cryoelectron microscopy) of a microtubule, courtesy Ken Downing, Lawrence Berkeley National Laboratory. E. Nogales, M. Whittaker, R.A. Milligan and K.H. Downing. 1999.

structure of a single microtubule, as well as the individual tubulin protein in its alpha and beta forms.

Chemical Imaging Technique Involved

Light microscopy (confocal, multiphoton), X-ray microscopy, and electron microscopy are the mainstays of imaging instrumentation for biological imaging.

Insights Obtained Using Chemical Imaging

Observation of cellular and molecular details in living cells is possible. Structural and chemical information about cellular activities within single cells can be obtained.

Imaging Limitations

Better resolving power (at or near atomic resolution) is needed in biological imaging instruments. Imaging of discrete chemical environments in living systems remains out of reach.

Opportunities for Imaging Development

Improved optics sources, more robust fluorescent probes, molecular tags that are X-ray excitable, and more sensitive detectors will contribute much to the ability to image molecular interactions within cells at desired resolutions.

Case Study 7: Crystals and Their Structure (Data Mining and Storage)

Whether one obtains a crystal structure of material using an experimental imaging technique or a simulation, a potential problem is the large amount of information that must be stored if one chooses to retain the position of every atom for a single image. One approach to solving this complexity is to retain only the positions of those atoms that deviate significantly from the corresponding regular lattice (e.g., defects such as dislocations or disclinations in a crystal). Figure 2.14 illustrates the substantial reduction of extraneous information when only the dislocations are included. Not only is it easier to see the important features (dislocations that affect the material's optical properties), but equally important, this represents a substantial reduction in the memory storage needed to retain all of the atomic positions in a trajectory. Such a figure represents a substantial reduction in the memory storage that would be exaggerated in a trajectory. It also allows quick assessment of the degree of dislocation under the specified conditions. Although this simulation was obtained using an atomistic molecular dynamics (MD) approach, similar information can be obtained with reduced-dimensional approaches (multiscaled), such as the phase-field model of Goldenfeld in which the dynamics are carried out at mesoscopic length scales.[21,22,23] Figure 2.15 illustrates this in the case of a two-dimensional crystal. A key advantage of such a representation is that its dynamical structures can be computed so quickly that there is no need to store them; rather, one need only store the results by way of the parameterizations of the model and its parameterization for a particular material. Equally important, both sets of visual images allow the degree of heterogeneity in the samples to be seen and provide a guide to the experimentalist in harvesting information from specific samples.

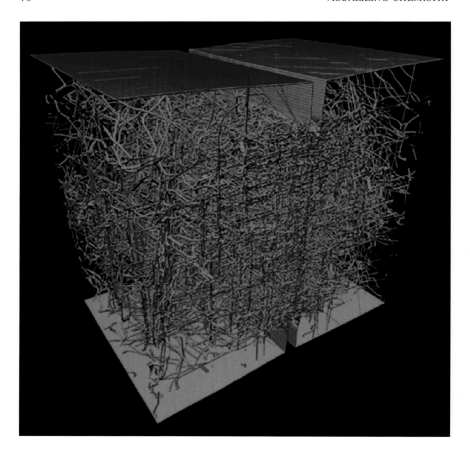

FIGURE 2.14 Two cracked tips (under stress) are pushed onto each other in an atomistic MD simulation. Only the microstructure of dislocations (comprised by a tiny percentage of the atoms) is displayed in green; all other atoms are hidden from view.
SOURCE: Yip, S. 2003. Synergistic science. *Nat. Mater.* 2:3-5; de Koning, M., A. Antonelli, and S. Yip. 2001. Single-simulation determination of phase boundaries: A dynamic Clausius-Clapeyron integration method. *J. Chem. Phys.* 115:11025–11035.

Chemical Imaging Technique Involved

• Reductionist or projective approaches to reduce large datasets (or systems) at the atomic length scale to important components at mesoscopic or macroscopic length scales

• Multiscaling and first-principles computational approaches to predict the structure and dynamics of materials given their atomistic or molecular composition

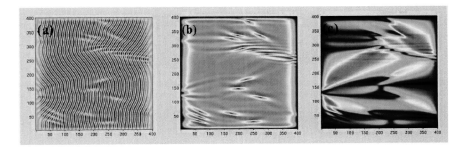

FIGURE 2.15 Comparison among different views of a snapshot of a subsystem satisfying the Swift-Hohenberg equation—a simplified model of convection in the absence of mean flow. Panel (a) shows the detailed flow directions, whereas panels (b) and (c) exhibit the amplitude and phase, respectively. The latter are slowly varying and allow for easier identification of the grain boundaries of the flow.
SOURCE: Courtesy of Nigel Goldenfeld's lecture, available on-line at *http:// guava.physics.uiuc.edu/~nigel/articles/RG/Patterns,%20universality%20and%20 computational%20algorithms.pdf* .

Insights Obtained Using Chemical Imaging

Using theoretical and computational techniques, one can identify the mesoscopic structures leading to a requisite function. Once identified, these structural motifs can be used to guide experimental chemical imaging probes.

Imaging Limitations

Computers with faster processors, larger random access memory (RAM), larger disks, and better communications bandwidth are needed. In addition, new computational codes capable of easily multiscaling structures from angstroms to meters are required. The expected increases in computing power available at high-performance computing sites as well as that available on a user's workstation will clearly make some calculations more accessible in the future. However, the major obstacles that have to be overcome lie primarily in the development of chemical theory, algorithms, and computer software. That is, the primary problem is the construction of static and dynamic structures that are simultaneously correct at resolutions ranging from the nanometer to the meter scale. This problem will not be solved easily by computing power alone.

Opportunities for Imaging Development

Development of fast, accurate, and user-friendly computer codes capable of multiscaling is necessary.

Case Study 8: Molecular Motors

Although biology abounds with amazingly complex molecular systems, perhaps the most astounding of these are the variety of molecular motors that perform the nanoscale mechanical work of living systems. Converting the chemical energy of adenosine 5^1-triphosphate (ATP) to mechanical work, these motors turn, or step, or induce enzymatic reactions. A variety of imaging techniques have been applied to the investigation of molecular motors. The most revealing have been performed at the single-molecule level. Indeed, the study of molecular motors is one of the most cited successes of single-molecule imaging. Single-molecule fluorescence has been used to visualize molecules moving or being moved by molecular motors in a field. Scanning probe spectroscopies have been adapted to measure the forces and mechanical parameters of motor function. Piconewton forces and nanometer movements have been measured using these techniques, opening a world of nanomechanics that previously was entirely unknown.

Chemical Imaging Technique(s) Involved

The techniques involved are single-molecule spectroscopy including optical (fluorescence) methods and single-molecule mechanical manipulation including scanning probe techniques (e.g., force measurements).

Insights Obtained Using Chemical Imaging

Chemical imaging at the single-molecule level has led to new understanding of mechanisms of molecular motors (force, torque, etc.).

Imaging Limitations

Imaging methods used at the single-molecule level can be applied only to selected molecules or molecular machines, primarily outside a living cell. It is important to study how individual molecules work together, ultimately in living cells.

Opportunities for Imaging Development

Single-molecule imaging techniques with improved temporal and spatial resolution have to be developed. Of particular importance is the ability to follow the dynamic activities of a single molecule, such as movements, structural changes, and catalytic functions. Future single-molecule studies both in vitro and in vivo will generate new knowledge of the working of molecular motors and other macromolecule machines and uncover mysteries in living systems (Figures 2.16 and 2.17).

FIGURE 2.16 A stylized cartoon of myosin V "walking" along an actin filament. Myosin V has the function of carrying cargo while walking along an actin tightrope and progressing in 37 nm steps. This movement has been observed using total internal reflection fluorescence (TIRF) spectroscopy of individual myosin molecules.

SOURCE: Reprinted with permission from 2003. *Science* (cover), 300(5628), based on Yildiz, A., J.N. Forkey, S.A. McKinney, T. Ha, Y.E. Goldman, and P.R. Selvin. 2003. Myosin V walks hand-over-hand: Single fluorophore imaging with 1.5-nm localization. *Science* 300:2061-2065. Copyright 2003 AAAS.

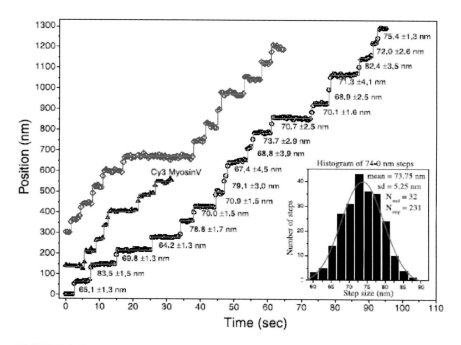

FIGURE 2.17 Results of total internal reflection fluorescence (TIRF) measurements in which the discrete and processive stepping action of myosin can be seen clearly.

SOURCE: Reprinted with permission from Yildiz, A., J.N. Forkey, S.A. McKinney, T. Ha, Y.E. Goldman, and P.R. Selvin..2003. Myosin V walks hand-over-hand: Single fluorophore imaging with 1.5-nm localization. *Science* 300:2061-2065. Copyright 2003 AAAS.

Case Study 9: Reverse Imaging

In addition to using imaging as a technique to obtain spatially and temporally patterned chemical data from a sample, one can also pattern chemical reactions in space and time using similar methods. Figure 2.18 demonstrates an example in which multiphoton scanning with ultrafast laser pulses was used to polymerize a photoresist resin with about 120 nm spatial resolution in three dimensions.

Photopolymerization is only one way in which the pattern and time course of chemical reactions can be controlled using imaging instruments. Atoms can be moved around on surfaces, specific genes can be turned on in one cell and not in a neighboring cell, and large arrays of heteropolymers (DNA, protein, etc.) can be synthesized on surfaces in which the chemical identity of each molecule at each position is distinct and known. (See Chapter 3 for further details.)

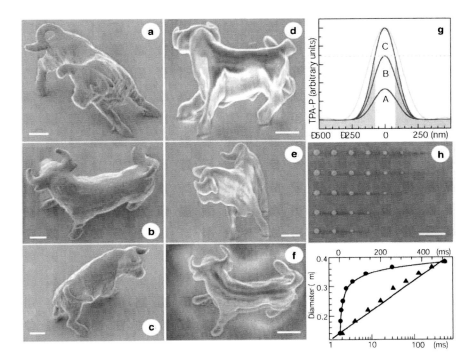

FIGURE 2.18 Example of reverse imaging in which multiphoton scanning with ultrafast laser pulses was used to polymerize a photoresist resin with about 120 nm spatial resolution. In panels a-f and h, the white bar is 2 microns.
SOURCE: Kawata, S., H.-B. Sun, T. Tanaka, and K. Takada. 2001. Finer features for microdevices. *Nature* 412:697-698.

Chemical Imaging Technique(s) Involved

Multiphoton microscopy has been used to initiate photopolymerization of a photoresist material in three dimensions with resolution in the hundred-nanometer range.

Insights Obtained Using Chemical Imaging

It is clearly possible to use chemical imaging not only to observe the structure and dynamics of chemical systems, but also to manipulate them at high resolution. This example presents a paradigm for the use of imaging to create much more complex chemical systems, patterned in three dimensions with extraordinary resolution.

Imaging Limitations

The wavelength of the light source employed dictates the fabrication resolution. In addition, in this case only a single chemical species, a photopolymer, is being manipulated. The potential exists for much more complex patterned chemical fabrications.

Opportunities for Imaging Development

There is no reason that chemical imaging in general cannot be turned on its head and used to manipulate chemical systems rather than just observe them. Because identifying features of chemistry can be observed in the imaging process shows that the probes used perturb this chemistry and that this perturbation can be patterned and controlled in both time and space. Considerably more could be done with high-throughput photopatterning of complex chemical systems. In addition, other techniques, such as control of individual magnetic particles in three dimensions with applied magnetic fields, should provide new vistas for fabrication and analysis as well as new opportunities for drug delivery in clinical settings.

Case Study 10: Terahertz Imaging for Electromagnetic Materials Research

One area of imaging spectroscopy that has attracted considerable attention recently is terahertz (THz)[24] radiation research. THz imaging is currently being touted in security- and defense-related applications, such as airport passenger and mailroom package screening. However, this case study focuses on the potential of time-resolved THz spectroscopy (TRTS). The THz frequency range spans the region between about 3 cm^{-1} (0.1 THz) to about 300 cm^{-1} (10 THz).[25] The radiation source may be generated from either continuous wave or short-pulsed lasers; the latter source of radiation allows TRTS studies to take place with subpicosecond temporal resolution.

THz spectroscopy was born from research efforts to produce and detect ultrashort electrical currents as they traveled down a transmission line.[26] In 1988-1989, it was discovered that electromagnetic radiation pulses produced by time-varying current could be propagated through free space and picked up by a detector.[27] By placing a sample between a THz source and detector, one could measure the differences in radiation pulses due to scattering or absorption by the sample to understand its chemical properties.

TRTS has numerous applications in materials science, chemistry, and biological research. One such application is in the study of charge transport in titanium dioxide (TiO$_2$), a material that is used in photovoltaic and photocatalytic systems. Scientists are currently studying electron transport in TiO$_2$ in order to better understand its photosensitive properties and engineer a more efficient surface for harnessing solar energy. Figure 2.19 shows a Grätzel solar cell, which

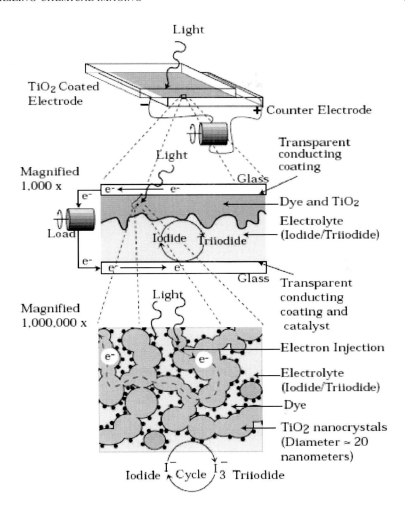

FIGURE 2.19 A schematic of the Grätzel solar cell.
SOURCE: Smestad, G.P.; M. Grätzel. 1998. Demonstrating electron transfer and nano-technology: A natural dye-sensitized nanocrystalline energy converter. *J. Chem. Educ.* 75: 752-756.

utilizes dye-sensitized TiO_2. Figure 2.20 shows the general scheme of dye sensi-tization of TiO_2. The photon energy of sunlight is not strong enough to excite an electron from the TiO_2 valence band to the conduction band in bulk; as a result, the surface of the TiO_2 film on a photovoltaic device is coated with a monolayer of a charge-transfer dye in order to photoexcite dye molecules that then inject

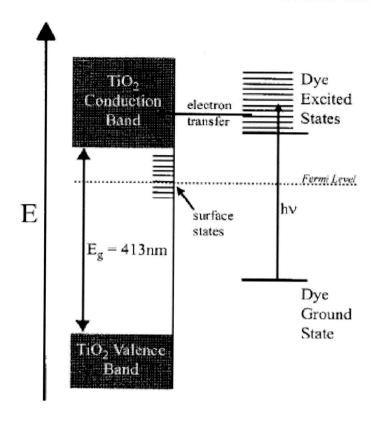

FIGURE 2.20 A schematic of the dye sensitization of TiO_2.
SOURCE: Beard, M.C., G.M. Turner, and C.A. Schmuttenmaer. 2002. Terahertz
spectroscopy. *J. Phys. Chem. B* 106:7146-7159.

electrons into the TiO_2 semiconductor.[28] TRTS can be used to dynamically
measure the mobilized electrons on a picosecond time scale within the conduction band without being affected by the dye molecules.[29]

In studies of TiO_2 conduction, TRTS has several advantages over fluorescence and other optical methods of spectroscopy. For example, one such advantage is that assumptions about electron behavior need not be made to analyze
spectra obtained through THz spectroscopy. Characterization of photoinjected
electron dynamics in dye-sensitized TiO_2 (Figure 2.20) has previously been
performed using the mid-infrared region of the spectrum.[30] However, in these
studies, it must be assumed that electron behavior follows the Drude model[31] to
account for transient infrared absorption of electrons. THz spectroscopy allows

complex conductivity responses to be obtained without assuming any prior model for electron behavior. In fact, TRTS revealed that the charge carriers significantly deviated from Drude behavior in colloidal, sintered TiO_2.[32] Furthermore, scattering responses have the greatest variation within the terahertz spectrum; thus a great deal of information regarding the dynamics of electron mobility may be obtained in this region. Finally, TRTS may be carried out at subpicosecond time resolution in order to follow the ultrafast dynamics of electron transfer within this system. Ultimately, the advantages of using TRTS to examine semiconductor materials may also be applied to spectroscopic methods for biological and medical imaging purposes.

Chemical Imaging Technique(s) Involved

Terahertz spectroscopy uses continuous wave (CW) and short pulsed laser excitation in the spectrum region between infrared and microwave frequencies. Pulsed laser excitation using pulse widths in the range of 10-100 femtoseconds has enabled the use of time-resolved terahertz spectroscopy, which is capable of capturing dynamic information at subpicosecond time scales.

Insights Obtained Using Chemical Imaging

Time-resolved terahertz imaging is capable of providing information about the dynamics of chemical reactions in materials science, chemistry, and biology.

Imaging Limitations

There exists a need for high-power pulsed CW radiation sources to enable fast switching times and high repetition rates for electromagnetic resonance experiments. In addition, commercial development of THz sources is needed so that this technology can be made more widely available to the research community. Furthermore, current detectors for THz spectroscopy have high cooling requirements to minimize noise in spectral data; further developments are needed to provide inexpensive and user-friendly detector options.

Opportunities for Imaging Development

Terahertz imaging offers the possibility of understanding complex reactions in which the chemical state of the sample under study changes with time. An extension of this ability is the control of chemical reactions in a highly specific manner; this will require the manipulation and channeling of the energy in a system such that the possible outcomes (degrees of freedom) are narrowed to those that one desires.

CONCLUSION

As demonstrated by the case studies presented in this chapter, chemical imaging has a wide variety of applications that have relevance to almost every facet of our daily lives. These applications range from medical diagnosis and treatment to the study and design of material properties in novel products. To continue receiving benefits from these technologies, sustained efforts are needed to facilitate understanding and manipulation of complex chemical structures and processes. Chemical imaging offers a means by which this can be accomplished by allowing the acquisition of direct, observable information about the nature of these chemistries.

NOTES AND REFERENCES

1. Kresge, C.T., M.E. Leonowicz, W.J.Roth, J.C. Vartuli, and J.S. Beck. 1992. Ordered mesoporous molecular sieves synthesized by a liquid-crystal template mechanism. *Nature*, 359:710-712.

2. (a) Huh, S., J.W. Wiench, B.G. Trewyn, M. Pruski, and V.S.-Y. Lin. 2003. Tuning of particle morphology and pore properties in mesoporous silicas with multiple organic functional groups. *Chem. Comm.* 2364-2365.

(b) Huh, S., J.W. Wiench, J.-C. Yoo, M. Pruski, and V. S.-Y. Lin. 2003. Organic functionalization and morphology control of mesoporous silicas via a co-condensation synthesis method. *Chem. Mater.* 15:4247-4256.

3. (a) Lin, V. S.-Y., D.R. Radu, M.-K. Han, W. Deng, S. Kuroki, B.H. Shanks, and M. Pruski. 2002. Oxidative polymerization of 1,4-diethynylbenzene into highly conjugated poly(phenylene butadiynylene) within the channels of surface-functionalized mesoporous silica and alumina materials. *J. Am. Chem. Soc.* 124:9040-9041.

(b) Huh, S., H.-T. Chen, J.W. Wiench, M. Pruski, and V.S.-Y. Lin. 2005. Cooperative catalysis by general acid and base bifunctionalized mesoporous silica nanospheres. *Angew. Chem. Int. Ed.* 44:1826-1830.

4. (a) Lai, C.-Y., B.G. Trewyn, D.M. Jeftinija, K. Jeftinija, S. Xu, S. Jeftinija, and V.S.-Y. Lin. 2003. A mesoporous silica nanosphere-based carrier system with chemically removable CdS nanoparticle caps for stimuli-responsive controlled release of neurotransmitters and drug molecules. *J. Am. Chem. Soc.* 125:4451-4459.

(b) Radu, D.R., C.-Y. Lai, K. Jeftinija, E.W. Rowe, S. Jeftinija, and V.S.-Y. Lin, 2004. A polyamidoamine dendrimer-capped mesoporous silica nanosphere-based gene transfection reagent. *J. Am. Chem. Soc.* 126:13216-13217.

5. Chang, H.- L., C. -M. Chun, I.A. Aksay, and W. -H. Shih. 1999. Conversion of fly ash into mesoporous aluminosilicate. *Ind. Eng. Chem. Res.* 38:973-977.

6. Ciuparu, D., R. F. Klie, Y. Zhu, and L. Pfefferle. 2004. Synthesis of pure boron single-wall nanotubes. *J. Phys. Chem. B* 108:3967-3969.

7. Barbara, P.F., A.J. Gesquiere, S.-J.Park, Y.J. Lee. 2005. Single-molecule spectroscopy of conjugated polymers. *Acc. Chem. Res.* 38:602-610.

8. de Boer, B., U. Stalmach, P.F. van Hutten, C. Melzer, V.V. Krasnikov, G. Hadziioannou. 2001. Supramolecular self-assembly and opto-electronic properties of semiconducting block copolymers. *Polymer* 42:9097-9109.

9. Gesquiere, A., M.M.S. Abdel-Mottaleb, S. De Feyter, F.C. De Schryver, F. Schoonbeek, J. van Esch, R.M. Kellogg, B.L. Feringa, A. Calderone, R. Lazzaroni, and J.L. Bredas. 2000. Molecular organization of bis-urea substituted thiophene derivatives at the liquid/solid interface studied by scanning tunneling microscopy. *Langmuir* 16:10385-10391.

10. Teetsov, J.A., and D.A. Vanden Bout. 2001. Imaging molecular and nanoscale order in conjugated polymer thin films with near-field scanning optical microscopy. *J. Am. Chem. Soc.* 123:3605-3606.

11. Wong, K.F., M.S. Skaf, C.-Y. Yang, P.J. Rossky, B. Bagchi, D. Hu, J. Yu, and P.F. Barbara. 2001. Structural and electronic characterization of chemical and conformational defects in conjugated polymers. *J. Phys. Chem. B* 105:6103-6107.

12. Schaller, R.D., L.F. Lee, J.C. Johnson, L.H. Haber, R.J. Saykally, J. Vieceli, I. Benjamin, T.-O. Nguyen, and B.J. Schwartz. 2002. The nature of interchain excitations in conjugated polymers: Spatially varying interfacial solvatochromism of annealed MEH-PPV films studied by near-field scanning optical microscopy (NSOM). *J. Phys. Chem. B* 106:9496-9506.

13. Barbara, P.F., A.J. Gesquiere, S.-J. Park, and Y.J. Lee. 2005. Single-molecule spectroscopy of conjugated polymers. *Acc. Chem. Res.* 38:602-610.

14. McNeill, J.D., D.Y. Kim, Z. Yu, D.B. O'Connor, and P.F. Barbara. 2004. Near field spectroscopic investigation of fluorescence quenching by charge carriers in pentacene-doped tetracene. *J. Phys. Chem. B* 108:11368-11374.

15. Muller, E.M., and J.A. Marohn. 2005. Microscopic evidence for spatially inhomogeneous charge trapping in pentacene. *Advanced Materials* 17:1410-1414.

16. Barbara, P.F., A.J. Gesquiere, S.-J. Park, and Y.J. Lee. 2005. Single-molecule spectroscopy of conjugated polymers. *Acc. Chem. Res.* 38:602-610.

17. Nonsense suppressors are produced by tethering a nonnatural amino acid to a stop (or "nonsense") anticodon in tRNA. As a result, the stop codon in an mRNA sequence is converted from a protein synthesis termination site to a site at which the nonnatural amino acid may specifically be inserted. DNA base substitutions that correspond to the stop anticodon of tRNA may thus be made in order to specifically incorporate nonnatural amino acids into proteins.

18. Wang, L., and P.G. Schultz. 2005. Expanding the genetic code. *Angew. Chem. Int. Ed.* 44:34-66.

19. Monahan, S.L., H.A. Lester, and D.A. Dougherty. 2003. Site-specific incorporation of unnatural amino acids into receptors expressed in mammalian cells. *Chem. Biol.* 10:573-580.

20. Information and images for this case study are provided courtesy of the Cancer Research Microscopy Facility, University of New Mexico Hospital; the W.M. Keck foundation; and the following individuals: Anthony L. Garcia, Linnea K. Ista, Dimiter N. Petsev, Michael J. O'Brien, Paul Bisong, Andrea A. Mammoli, Steven R.J. Brueck, and Gabriel P. Lopez.

21. Jeong, J.-H., N. Goldenfeld, and J. Dantzig. 2001. Phase field model for three dimensional dendritic growth with fluid flow. *Phys. Rev. E* 64:041602.

22. Vetsigian, K., and N. Goldenfeld. 2003. Computationally efficient phase-field models with interface kinetics. *Phys. Rev. E* 68:60601.

23. Goldenfeld, N., B.P. Athreya, and J.A. Dantzig. 2005. Renormalization group approach to multiscale simulation of polycrystalline materials using the phase field crystal model. *Phys. Rev. E* 72:1-4.

24. THz spectroscopy is also known as far-infrared (FIR) spectroscopy.

25. 1 THz is equivalent to 33.33 cm^{-1} (wavenumbers), 0.004 eV photon energy, or 300 μm wavelength.

26. (a) Schmuttenmaer, C.A. 2004. Exploring dynamics in the far-infrared with terahertz spectroscopy. *Chem. Rev.* 104:1759-1779.

(b) Beard, M.C., G.M. Turner, and C.A. Schmuttenmaer. 2002. Terahertz spectroscopy. *J. Phys. Chem. B* 106:7146-7159.

27. (a) Smith, P.R., D.H. Auston, and M.C. Nuss. 1988. Subpicosecond photoconducting dipole antennas. *IEEE J. Quantum Electron.* 24:255-260.

(b) Fattinger, C., and D. Grischkowsky. 1989. Terahertz beams. *Appl. Phys. Lett.* 54:490-492.

28. (a) O'Regan, B., and M. Grätzel. 1991. A low-cost, high-efficiency solar cell based on dye-sensitized colloidal TiO$_2$ films. *Nature* 353:737-740.

(b) Beard, M.C., G.M. Turner, and C.A. Schmuttenmaer. 2002. Terahertz spectroscopy. *J. Phys. Chem. B* 106:7146-7159.

29. Turner, G.M., M.C. Beard, and C.A. Schmuttenmaer. 2002. Carrier localization and cooling in dye-sensitized nanocrystalline titanium dioxide. *J. Phys. Chem. B* 106:11716-11719.

30. (a) Heimer, T.A., and E.J. Heilweil. 1997. Direct time-resolved infrared measurement of electron injection in dye-sensitized titanium dioxide films. *J. Phys. Chem. B* 101:10990-10993.

(b) Gosh, H.N., J.B. Asbury, and T. Lian. 1998. Direct observation of ultrafast electron injection from coumarin 343 in TiO_2 nanoparticles by femtosecond infrared spectroscopy. *J. Phys. Chem. B* 102:6482-6486.

31. The Drude model applies the kinetic theory of gases to metal conduction. It describes valence electrons as charged spheres that move through a "soup" of stationary metallic ions with finite chance for scattering.

32. Beard, M.C., G.M. Turner, and C.A. Schmuttenmaer. 2002. Terahertz spectroscopy. *J. Phys. Chem. B* 106:7146-7159.

3

Imaging Techniques:
State of the Art and Future Potential

The case studies in Chapter 2 underscore the power of chemical imaging to provide insights into a wide variety of problems in the chemical sciences. In this chapter,[1] the current capabilities of chemical imaging are examined in detail, as are areas in which basic improvements in imaging capabilities are needed. However, the chapter is not intended to be an exhaustive review of all chemical imaging techniques. It is assumed that the reader has a basic knowledge of the imaging techniques described. The objective of this chapter is to provide an overview of the state of the art in chemical imaging and to identify those areas that would most likely provide breakthroughs.

The imaging techniques described are divided into three main categories. In addition, a section on image processing and computation—which has bearing on virtually all chemical imaging techniques—is also included:

- Optical imaging (Raman, infrared [IR], and fluorescence) and magnetic resonance
  - Electron microscopy, X-rays, ions, neutrons
  - Proximal probe (force microscopy, near field, field enhancement)
  - Processing analysis and computation

OPTICAL IMAGING AND MAGNETIC RESONANCE

Imaging techniques that utilize low-energy resonant phenomena (electronic, vibrational, or nuclear) to probe the structure and dynamics of molecules, molecular complexes, or higher-order chemical systems differ from approaches using higher-energy radiation (X-rays, electrons, etc.) in that they are largely nondestructive and can be performed under in vivo or in situ conditions, even with soft matter. However, these techniques lack the inherent spatial resolution of the higher-energy approaches.

Although similar in these respects, magnetic resonance and optical spectroscopy (electronic and vibrational spectroscopy) have different strengths and weaknesses. Magnetic resonance is the lowest-energy method and as such uses the longest-wavelength radiation. Exquisite detail in molecular structure can be defined due to the fact that atomic interactions can be measured. However, this detail about the atomic interactions is accompanied by a low inherent sensitivity, thus requiring extensive averaging over many molecules and limiting the inherent temporal and spatial resolution. In contrast, optical spectroscopy utilizes radiation at an energy level high enough to allow individual photons to be measured relatively easily with modern equipment at a detection sensitivity almost matched by the mammalian eye. As a result, imaging data are acquired at the sensitivity of individual molecules. The inherent temporal and spatial resolution is also increased proportionately, but the resonance itself is broad because environmental influences are not averaged out within the inherent time scale of interaction between the molecules and this frequency of radiation. As a result, the structural information content of optical spectra is considerably lower than that of magnetic resonance, particularly in the electronic region of the spectrum.

The long-term technical challenge is to extract the maximum possible information from each type of resonance, ultimately providing a detailed structural picture of the chemistry at the molecular level with the spatial resolution of individual molecules and a temporal resolution on the time scale of chemical bonding.

Nuclear Magnetic Resonance

Over the past 50 years, nuclear magnetic resonance (NMR) has grown into an essential tool for chemists in determining structures of newly synthesized compounds, for scientists interested in the structure of solids, and for biochemists in determining structure-function relationships in biomolecules. NMR also forms the basis for magnetic resonance imaging (MRI). The incredible breadth of NMR and its impact on chemical, biological, and medical sciences have created a vibrant and innovative community of scientists working to increase the scope and usefulness of NMR. Many books are dedicated to subsets of the techniques involved in NMR and MRI: thus, the goal here is to give a small taste of the types of information available and to point out areas in which progress would impact a large subset of NMR and MRI experiments. In addition, there is an equally rich field, which is not discussed explicitly, that applies electron spin resonance to many of the same problems to which NMR and MRI are applied.

Recent advances have pushed the limits of molecular structure determination, including applications of NMR to larger and larger molecules and new ways to enhance the detection limits of NMR. MRI has also undergone a major transition from a tool that provides primarily anatomical information to one that can measure a number of aspects of tissue function. Indeed, active areas of the human brain can now be mapped at unprecedented resolution using functional MRI. However, there is much room for improvement, and there are a number of fruitful areas for development. Higher-magnitude magnetic fields, more sensitive detection strategies, and an ever-growing list of MRI contrast agents will continue to expand the usefulness of NMR and MRI, rendering them essential in chemical imaging. This section provides a general outline of the present state of the art of NMR and MRI, describes some exciting new developments in the area, and finally points out some opportunities for future work that can impact NMR and MRI broadly.

Present State of the Art

Nuclear Magnetic Resonance Spectroscopy: Molecular Structure and Dynamics. NMR is the only tool that provides detailed three-dimensional information at angstrom (Å) resolution of molecules both in solution and in noncrystalline solids. NMR is thus important in imaging molecules not only for the organic chemist but also for materials scientists and biochemists. Its exquisite sensitivity to molecular structure is due to the ability to monitor interactions between atoms that report on structure and dynamics. Chemical shift and J-coupling information obtained from NMR is the result of specific chemical bonds and bond angles. Through-space interactions, such as dipole-dipole interactions, are sensitive to short range (1-5 Å) nonbonded information. Thus, rather than using diffraction of radiation as in X-ray crystallography, NMR builds up structures from a large number of specific interatomic distances and bond angles. Over the past 30 years, the development of complex multidimensional NMR experiments on molecules isotopically labeled with ^{15}N, ^{13}C, and ^{2}H has made routine the probing of detailed structures of molecules in solution up to a molecular weight of approximately 40,000. Similar developments in solid-state NMR now allow a number of structural constraints to be obtained for much larger molecules. The awarding of the Nobel Prize in chemistry in 1991 to Richard Ernst for his work in developing fundamental strategies in NMR and in 2002 to Kurt Wuthrich for his work in using NMR to solve protein structures testifies to the impact of NMR.[2]

In addition to structural information, dynamic information can also be obtained through NMR. Time scales of both fast (picoseconds) and slow (seconds and longer) processes can be followed. Slow processes such as chemical reactivity are probed by following a change in an NMR property such as chemical shift or transfer of magnetization from one spectral site to another. Detailed kinetic information can be extracted in well-established experiments. Faster processes influence the NMR spin relaxation properties, such as T_1 or T_2, with kinetic information linked to the specific structure being examined. Model-independent ways of analyzing relaxation data have enabled very efficient procedures for determining which parts of a molecule are more dynamic and over what time scales the fluctuations occur. Thus, NMR is unmatched in the detailed structural and dynamic information it offers.

The main limitation of NMR continues to be its relatively low sensitivity, requiring homogeneous (or heterogeneous mixtures with only a few components) samples of relatively high concentrations (e.g., a milliliter of 10 mM concentration) to be studied. Separation techniques such as high-performance liquid chromatography (HPLC) can be performed prior to NMR to help study complex mixtures, but the ability to obtain detailed structural information about complex mixtures that vary at high spatial resolution requires large gains in sensitivity. Three major directions are being pursued to increase sensitivity. First, higher-magnitude magnetic fields increase anywhere from linearly to quadratically in sensitivity with an increase in field strength, depending on the sample. Magnets with fields up to about 20 Tesla operating at 900 MHz frequencies are becoming available at a few dedicated research sites. A second pursuit has been the improvement of detectors for NMR. One such strategy that has become widely available over the past five years is cooling of the NMR detectors to reduce noise, which has increased sensitivity by a factor of 2 to 4. Work is progressing to miniaturize NMR detectors and use detector arrays to increase sensitivity and throughput. Furthermore, work is aimed at using innovative approaches to detect magnetic resonance signals, such as magnetic force microscopy,[3] which borrows concepts from near-field imaging, and other classes of detectors continue to be developed, such as superconducting quantum interference devices (SQUID) for NMR.[4]

A third approach to increase sensitivity is to increase the signal available from a molecule using hyperpolarization techniques. Indeed, hyperpolarization techniques are leading to large increases in sensitivity from 100- to 100,000-fold. Techniques to transfer polarization were pioneered by physicists such as Albert Overhauser, who was awarded the National Medal of Science in 1994 for his work predicting that electron spin polarization could be coupled to nuclear spin polarization, and Alfred Kastler, who was awarded the Noble Prize in physics in 1966 for his work demonstrating that optical pumping could lead to hyperpolarization. These techniques are now beginning to find widespread application. When samples are placed in the magnets typically used for NMR, at least a million spins are required to generate enough of a population difference between ground and excited states to give a signal. In practice, many more molecules are needed for a sufficient signal to be generated for detection. There is a class of techniques that rely on transferring polarization from molecules that have greater population differences to molecules that one would like to detect with NMR and in this way generate a larger population difference with much fewer spins. There are numerous ways to transfer polarization and increase signal. Three specific techniques that have found growing use are transfer of polarization from unpaired electrons in stable free radicals to nuclear spins,[5] laser-induced polarization of noble gases such as xenon and helium,[6] and chemical formation of molecules from parahydrogen that can be produced in a polarized state.[7] These hyperpolarization strategies are being used to increase sensitivity for application to a wide range of problems in physics, chemistry, biochemistry, and medical imaging.

In addition to increasing the sensitivity of NMR, much work is being done to improve the specificity and accuracy of information available from NMR. Perhaps this is most evident in work on biological macromolecules, which is an active area of development for NMR. An exciting recent example shows that partial orientation of molecules in solution greatly increases the strength of dipole-dipole interactions that are important for obtaining distance information. The strategy of partial alignment has led to structural information about molecules (such as proteins) at very high resolution and with very high accuracy.[8] There are also a variety of new NMR techniques to measure dynamics of complex molecules in solutions. In general, these techniques rely on measuring NMR relaxation times and interpreting them in the context of a model of the motion. Recent work measuring the relaxation time of deuterium has enabled the measurement of side chain motion of proteins in solution, with molecular weights up to about 100,000 daltons.[9] Indeed, a variety of sophisticated NMR pulse sequences enable motion to be analyzed on the picosecond through millisecond time scale. Development of these pulse sequences continues to be an active area of research. Finally, much of the information about structure and dynamics obtained in the solution state by NMR can also be obtained using solid-state NMR for molecules of much higher molecular weight. Detailed structural and dynamic information can be obtained even if the material being studied defies crystallization.[10] The exciting area of solid-state NMR is rapidly developing for determining structures of novel materials important for nanotechnology as well as for proteins that do not readily crystallize.

Magnetic Resonance Imaging: Noninvasive Measurement of Anatomy, Function, and Biochemistry. In 1974, Paul Lauterbur introduced a gradient field strategy to obtain images based on NMR. Today, MRI is being employed in more than 10 million scans per year in the United States and is thus having a great impact on the diagnosis and treatment of a wide variety of diseases. Its importance was recognized when the 2003 Nobel Prize in medicine was awarded to Drs. Lauterbur and Mansfield.[11] The basis for MRI is the change in chemical shift that an atom undergoes in an applied magnetic field. With proper calibration of the magnetic field gradient, a change in chemical shift can be related to a specific location—a process known as frequency encoding of spatial information. In addition, controlling the applied magnetic field gradients in combination with specific radio-frequency pulses to excite specific regions enables signals to come from these specific regions—a process known as slice selection. Finally, the time evolution of the NMR signal during a series of radio-frequency excitation pulses can be modulated by the chemical shift of the nucleus being detected. Because the chemical shift can be altered by applied magnetic field gradients during these evolution times, spatial information can be obtained—a process known as phase encoding. There is a wide variety of techniques that use innovative combinations of frequency encoding, slice selection, and phase encoding to generate images.

Any nucleus that can be detected by NMR can be imaged with MRI. The most widely used atom is the hydrogen in water because the high concentration of water enables high-resolution images and a large amount of information can be obtained about the environment of water from changes in its NMR relaxation times, T_1 and T_2. However, much work has been done detecting other nuclei such as ^{23}Na, ^{31}P, and compounds labeled with ^{13}C, to name a few. In most cases, MRI is performed on the hydrogen atoms in water and detects the single NMR peak from water. However, strategies referred to as spectroscopic imaging or chemical shift imaging enable a series of images to be obtained that represent every resonance in an NMR spectrum. In this way, images of complex metabolite distributions have been obtained and applied to get a metabolic fingerprint of normal and diseased tissue.

Interaction of the hydrogen on a water molecule (or any other NMR active nucleus) with an applied magnetic field gradient enables MRI to create images at much higher resolution than the wavelength of the applied radiation, leading to images with resolutions in the range of 0.2-3 mm in humans and as low as 0.05 mm in animals. With small samples at high magnetic fields, resolution as low as a few microns has been achieved. This is a key factor in the ability of MRI to obtain high resolution of tissues nondestructively using long-wavelength, and thus low-energy, nonionizing radiation. The second reason behind the usefulness of MRI is the remarkable degree of specificity and sensitivity to disease. Water reports on changes in its environment, and the relaxation times of water are sensitive to specific tissues, enabling unparalleled anatomical information to be obtained from soft tissues in the body. In addition, spectroscopic imaging gives information about a large range of metabolites that can be affected early in disease processes. The largest application of MRI has been to biomedical problems, but there is a growing list of problems from characterization of solids to understanding fluid flow in complex media that have been addressed with MRI. Indeed, funding to translate developments of MRI in the biomedical arena to other areas central to chemical imaging would have a major impact.

The past decade has seen a rapid growth in the use of MRI to obtain anatomical information and functional information about tissues. Strategies have been developed that enable MRI to generate images of flowing water, enabling angiography to be performed on the circulatory system. MRI can also be used to measure bulk flow of water, allowing regional blood flow to be measured from a number of tissues. NMR has been used for decades to measure the magnitude and direction of molecular diffusion in solution, and it is possible to extend these techniques to MRI. Techniques for measuring regional blood flow and diffusion are having a major impact on assessing ischemic disease such as heart attacks and stroke. Indeed, at an early stage, diffusion and perfusion MRI can be used to decide therapeutic strategies for stroke victims. In addition, MRI can be sensitized to blood oxygenation levels to assess the degree of metabolic activity in a region of a tissue. When a region of the brain becomes active, the increases in blood flow and metabolism lead to changes in blood oxygenation that can be detected by MRI. This oxygenation-dependent, functional MRI contrast has revolutionized cognitive psychology and is leading to a detailed understanding of the regions of the brain that are responsible for complex cognitive functions.[12] Finally, NMR spectroscopy can be combined with MRI to generate detailed spectroscopic images of a range of metabolites. The entire range of functional MRI tools is poised to have a major impact on the diagnosis and management of disease.[13]

The Cutting Edge and Future Directions in NMR and MRI

Higher Magnetic Fields. The sensitivity of magnetic resonance increases with higher magnetic fields. Indeed, in the range where detector noise dominates, sensitivity increases as approximately the square of the increase in field. In practice, this is hard to realize, particularly because many samples of interest contribute noise, leading to an increase in sensitivity that is linearly proportional to magnetic field strength. Nonetheless, much interest has been focused on producing higher magnetic fields for NMR. Most of this work occurs in industry where fields as high as 20 Tesla (T) can be produced for routine analytical chemistry and biochemistry. In MRI, magnets up to 9.4 T that are large enough for humans are becoming available. These high fields should increase the resolution of MRI of hydrogen as well as be a great boost to MRI of nuclei less sensitive than hydrogen, such as ^{23}Na, ^{31}P, and ^{13}C. The cutting edge for development of high-field magnets is at the National Magnet Laboratory at the University of Florida, where magnets as high as 40 T are available for use.[14] In France, a new project is proceeding to increase the strength of magnetic fields available for MRI on humans to 12 T.[15] Transforming these exciting projects into commercially viable products would have widespread impact and enable the development of new technologies that allow even higher magnetic fields to be created. This major challenge is in need of creative thinking to move forward without the very great expenditures that these projects currently require. For example, with present magnet technology, significant space is required to house a high-strength magnet. Work to decrease the siting requirement of high-field magnets, for example by employing innovative designs for superconducting wire that can carry higher current densities, could decrease the size of magnets, enabling very high field NMR and MRI to transition from dedicated laboratories to widespread use.

There is some work indicating that NMR can become a more portable modality. For example, in the oil industry the NMR system is attached directly to the exploration drill to mine for petroleum sources. A generalization of this portability of NMR could lead to applications in a range of environmental studies as well as in medical contexts, where a handheld MRI device might be available to clinicians working far from a hospital's radiology department. Recently, the use of a SQUID detector has been demonstrated to lead to excellent NMR spectra at very low magnetic fields, pointing to the possibility of making NMR more portable.[16] Thus, there is much room for innovative work, both to enable higher magnetic fields and to make NMR more portable with lower magnetic fields.

Development of New MRI Detectors. Another important strategy for increasing sensitivity in NMR and MRI is the development of new detectors. For NMR, an increase in sensitivity from two- to fourfold has occurred by decreasing the temperature of the detector. These advances, using either high-temperature superconducting materials or traditional materials, are now being implemented widely. There have been similar sensitivity gains in MRI due to the widespread availability of high magnetic fields (3-9 T) for human use and the development of parallel detector arrays. Five years ago, for example, an effective scan of a human head was achieved with an MRI detector containing only one element. Today detectors with 8 to 32 elements are becoming common,[17] with preliminary data obtained from arrays with up to 90 elements. These arrays increase sensitivity from two- to fivefold and also enable MRI to be performed at much faster speeds.[18] When these arrays are dense enough for the coil noise to dominate over the sample noise, cooling arrays should increase the sensitivity of MRI further. The challenge is to insulate the detectors so that very cold temperatures can be achieved while keeping the detectors close to the body so that sensitivity gains can be realized. With the rapid increase in detector density, it is critical to develop strategies that enable miniaturization of the electronics necessary to perform MRI. A concerted effort to miniaturize NMR components not only will enable engineering of dense detector arrays, but also should increase the portability of NMR in general.

There is much to gain by focusing research efforts to increase sensitivity in NMR and MRI. At present, MRI on humans is performed at resolutions of about a millimeter, with recent results pushing these limits to about 300 microns. A factor of 100X gain in sensitivity would place MRI on the brink of detecting single cells in any organ within the human body. This would also enable chemical imaging for a larger variety of problems where the unmatched chemical sensitivity of NMR can be combined with the spatial resolution afforded by MRI. Research on other detector strategies besides those commonly used should be encouraged, for example developing SQUID detectors for NMR or other innovative approaches to detecting signals. Indeed, it is only the lack of sensitivity that at present limits widespread application of MRI as a chemical imaging tool to the full range of problems discussed throughout this report.

Increasing the NMR Signal with Hyperpolarization. A very promising avenue for increasing sensitivity in NMR and MRI is to increase the signal from the molecules being detected. The low radio-frequency energy used for NMR means that specific nuclei in molecules are as likely to be in the excited state as in the ground state. Signal detection is proportional to the population difference between the two states. Typically, it takes a million molecules to generate a larger ground state than excited state population. There are a number of ways to alter this population difference and polarize the sample to obtain more signals. As discussed previously, increasing the magnetic field for NMR and MRI is one way to achieve incremental gains. Another alternative is to decrease the temperature, which is useful only if the sample is amenable to lower temperatures. A final and very dramatic way is to couple the nuclear spins being detected by NMR to other spins with a higher polarization. As mentioned earlier, transferring polarization from electrons, optically pumping to achieve increased nuclear polarization of noble gases, and using parahydrogen have all been successful in increasing the signal by as much as 100- to 100,000-fold. For example, so-called dynamic nuclear polarization experiments coupling a stable free radical to NMR-detectable nuclei have demonstrated great gains in sensitivity for solid-state NMR, enabling experiments that would ordinarily last days to be performed in minutes.[19] Furthermore, clever strategies allow the solid to be thawed to a liquid and prepared in a manner such that it can be injected, which enables hyperpolarization to be used in vivo for MRI. Hyperpolarized MRI of ^{13}C-labeled compounds has been shown to increase sensitivity more than 100,000-fold; this offers exciting possibilities to trace specific metabolic pathways to identify diseases such as cancer.[20] One major drawback is that these techniques cannot be applied generally to all molecules. Optical pumping of the noble gases xenon and helium can also lead to very large gains in sensitivity. Recent work has demonstrated the potential for producing biosensors from optical-pumped xenon to enable detection to about 200 nM.[21] Hyperpolarized noble gases are also finding increasing use for MRI of the air spaces in lungs.[22]

A major shortcoming of these hyperpolarization studies is that they are applicable to only a few molecules. Generation of new materials optimized for hyperpolarization is very important to enable a large range of molecules to be hyperpolarized. Another major limitation is that the hyperpolarized signal lasts for a time defined by the nuclear spin lattice relaxation time. In the molecules being developed this means that the increased signal lasts for about a minute. Innovative approaches to making the best use of the polarization while it lasts and procedures for replenishing the signal are critical to a broader range of application. Ideally, a new generation of physicists, chemists, and biochemists would be trained to conduct this truly interdisciplinary work.

Detection of Single Spins with Scanning Force Magnetic Resonance. Within the last year the detection of a single electron spin was accomplished with a scanning magnetic resonance experiment using cantilevers similar to those used for scanning force microscopy.[23] This was the culmination of many years of progress to detect increasingly fewer electron or nuclear spins using the magnetic resonance phenomenon. The experiment relied on measuring the force generated when the electron spin orientation was flipped by application of the appropriate radio frequency in a magnetic field. Because the electron spin is 1,000 times stronger than a nuclear spin, this result opens the possibility of detection of single nuclei and thus single-molecule detection by magnetic resonance. As a result, one can envision the use of a small cantilever to scan a molecule or molecular assembly to determine its detailed chemical composition and three-dimensional structure. Such an advance will take years of development to realize and requires advances similar to those needed in other scanning near-field imaging techniques, including (1) the development of more sensitive cantilever strategies to measure increasingly smaller forces and (2) a deeper theoretical understanding of single-molecule behavior with respect to magnetic resonance.

Quantitative Understanding of Chemical Shifts. A great triumph for NMR has been the ability to obtain detailed three-dimensional information from molecules with weights up to about 40,000 grams per mole with accuracy to a few angstroms. It is well known that NMR chemical shifts are sensitive to very small bond length and bond angle changes and can thus probe chemical potentials at very short distances. This is due to the exquisite sensitivity of nuclear spins to their electronic environment. One of the great challenges of modern chemistry is to develop quantum mechanical calculations that can predict chemical interactions and chemical reactions of large molecules. A great hurdle to this work is developing analytical tools that can measure potential changes over short distances. Analysis of the chemical shift of nuclei is one of the few techniques that can probe these potentials over short distances. Thus, a critical frontier in work in NMR is to develop computational approaches that enable prediction of chemical shifts in large molecules. Indeed, if this work is successful it will be possible to determine molecular structures of very complex molecules in a time-efficient manner to an unprecedented level of resolution.

Novel Contrast Agents for MRI. Contrast agents have played an important role in the development of MRI. For example, simple gadolinium chelates are critical for the usefulness of MRI in detecting brain tumors, performing angiography, and measuring regional blood flow and metabolism. With the rapid developments in molecular genetics identifying a large number of potential indicators of disease and therapeutic targets, there is increasing interest in developing MRI contrast agents that are specific for particular cells, molecules, or biochemical processes. This emerging area of molecular imaging depends on the marriage of (1) chemical synthesis of new labels to add specificity to the agent and (2) MRI acquisition and processing to optimize strategies to detect these new agents. Recent work has demonstrated that MRI can be used to specifically target cell surface molecules, image gene expression, detect enzymatic reactions, and follow the migration of cells in intact organs.[24] These developments are a long way from routine clinical use, and the realization of this potential will take the concerted efforts of a multidisciplinary team of chemists, molecular biologists, radiologists, and MRI physicists. Particularly lacking are chemists with a commitment to work in this highly multidisciplinary area. Furthermore, the general strategies being offered are applicable to a broad range of problems outside the field of medicine, such as detection of sparse molecules of environmental interest or characterization of complex materials. Funding to translate developments in the biomedical area to broader use in chemical imaging would have a great impact.

Conclusions

NMR and MRI represent mature technologies that have widespread impact on the materials, chemical, biochemical, and medical fields. Recent results in determining the structures of key biological macromolecules and the transformation of the cognitive sciences due to functional MRI exemplify this tremendous influence. Despite these achievements, there is much progress yet to be made. Research aimed at improving magnet technology to achieve higher field strengths in smaller footprints will advance the sensitivity and applicability of NMR. Developments to miniaturize NMR electronics will greatly aid the rapid progress in parallel detection for MRI and increase the portability of NMR. Investment in the exciting area of hyperpolarization has an excellent chance to greatly increase the sensitivity and applicability of NMR and MRI. Investments in the theoretical aspects of NMR, especially those that enable the prediction of structural information from chemical shifts and the optimization of approaches to increase sensitivity using hyperpolarization, will pay large dividends. Finally, funding toward development of new materials can impact NMR on many levels. New superconducting materials can impact magnet and detector design, and new approaches to generating sensitive cantilevers will usher in the era of single-molecule detection by magnetic resonance. A new generation of chemists can impact NMR and MRI research by focusing on the development of new molecules amenable to hyperpolarization strategies as well as new contrast agents to contribute to the rapidly growing field of molecular imaging. Funding mechanisms that can lead to faster translation of developments made in the biomedical area to other areas of chemical imaging should be pursued. It is clear that in the coming years, NMR and MRI will continue to expand rapidly and continue to be key tools for chemical imaging.

Vibrational Imaging

A vibrational spectrum provides something like a structural "fingerprint" of matter because it is characteristic of chemical bonds in a specific molecule. Therefore, imaging based on vibrational spectroscopic signatures, such as Raman scattering and IR absorption, provides a great deal of molecular structural information about the target under study.

Raman Scattering and Infrared Absorption Imaging

In particular, because of their high structural selectivity, Raman and IR imaging techniques also have the capability to monitor chemical structural changes that occur in chemical and physical processes. Both IR and Raman imaging techniques benefit from recent developments of array detectors, which allow the rapid collection of both spectral and positional data.

Infrared absorption spectroscopy is a straightforward technique for vibrational imaging. Infrared Fourier transform (FT) microscopy with scanning options allows "chemical mapping" with lateral resolution on the order of tens of microns. The integration of IR absorption spectroscopy into near-field scanning optical microscopy is a promising approach to in situ, nondestructive, high-spatial-resolution imaging, with applications in the chemical characterization of materials and nanotechnology that improve the spatial resolution of IR spectroscopy to 300- 500 nm attainable in the near field.[25] Due to the Raman effect, inelastically scattered light is shifted in wavelength relative to the excitation frequency by the characteristic molecular vibrational frequency of the probed material. Therefore, Raman scattering can be applied noninvasively under ambient conditions in almost every environment, including those in which water is present. Today, laser photons over a wide range of frequencies from the near-ultraviolet to the near-infrared region are used in Raman scattering studies, allowing selection of optimum excitation conditions for each sample. By choosing wavelengths that excite appropriate electronic transitions, resonance Raman imaging of selected components of a sample or parts of a molecule can be performed.

The range of excitation wavelengths has been extended to the near-infrared (NIR) region, in which background fluorescence is reduced and photoinduced degradation from the sample is diminished. Moreover, high-intensity diode lasers are easily available, making this region attractive for compact, low-cost Raman instrumentation. Furthermore, the development of low-noise, high-quantum-efficiency multichannel detectors (charge-coupled device, CCD), combined with high-throughput spectrographs and used in combination with holographic laser rejection filters, has led to high-sensitivity Raman spectrometers.

The main advantage of Raman spectroscopy is its capability to provide rich information about the molecular identities of the sample. Sophisticated data analysis techniques based on multivariate analysis have made it possible to exploit the full information content of Raman spectra and draw conclusions about the chemical composition of very complex systems such as biological materials.[26] The downside of vibrational imaging techniques comes from relatively small IR absorption cross sections and also from the extremely small cross section for Raman scattering (typically 10^{-30}-10^{-25} cm^2 per molecule), with the larger values occurring only under favorable resonance Raman conditions. The small cross sections result in very weak imaging signals. For comparison, effective fluorescence cross sections can reach about 10^{-16} cm^2 per molecule for high-quantum-yield fluorophores. On the other hand, particularly under ambient conditions, the amount of molecular structural information that can be obtained from fluorescence imaging is limited.

In terms of the high content of chemical structural information at desired spatial and temporal resolutions, Raman spectroscopy would be a very useful technique for chemical imaging. A disadvantage, however, in many applications of Raman imaging results from relatively poor signal-to-noise ratios due to the extremely small cross section of the Raman process, 12 to 14 orders of magnitude lower than fluorescence cross sections. New methodologies such as surface-enhanced Raman scattering and nonlinear Raman spectroscopy can be used to overcome this shortcoming.

Surface-Enhanced Raman Scattering. In the 1970s, a discovery that showed unexpectedly high Raman signals from pyridine on a rough silver electrode attracted considerable attention.[27] Within a few years, strongly enhanced Raman signals were verified for many different molecules, which had been attached to various "rough" metal surfaces; the effect was called "surface-enhanced Raman scattering" (SERS). The discovery of SERS showed promise to overcome the traditionally low sensitivity of Raman spectroscopy.

It soon turned out that enhanced Raman scattering signals are associated mainly with nanoscale roughness structures on the silver electrode, and similar and even stronger enhancement factors were observed both for small silver and gold colloidal particles in solution and for evaporated island films of silver and gold. Enhancement of Raman signals occurs due to resonances between the optical fields and the collective motion of the conduction electrons (surface plasmons) in metallic nanostructures. This resonance effect leads to strongly enhanced and spatially confined local optical fields in the close vicinity of metallic nanostructures where spectroscopy takes place, resulting in strongly enhanced Raman spectra. Enhancement of excitation and scattered field results in an increase in Raman scattering signal intensity equal to the fourth power of the field enhancement. In addition to this "electromagnetic" field enhancement effect, electronic interactions between the Raman molecule and the metal (e.g., charge transfer) can result in an increase of the Raman cross section itself; this is known as "chemical or electronic enhancement."[28] Strong enhancement factors that should be associated with a "chemical effect" have been observed recently for small metal clusters.[29] Although the Raman shifts, relative peak intensities, and line widths with SERS may differ slightly from those in normal Raman spectra due to a combination of the molecular interaction with the metal, high local confinement of the effect, and large field gradients, a SERS spectrum still provides a clear "fingerprint" of a molecule. Moreover, SERS is an analytical technique that can give information on surface and interface processes, such as charge-transfer processes at the nanoscale.[30]

The task of imaging single molecules while simultaneously identifying their chemical structures and monitoring structural changes poses a challenge that is of both basic scientific and practical interest in many fields. At present, SERS is the only way to detect a single molecule and simultaneously identify its chemical structure.

Extremely high SERS enhancement factors can bring effective Raman cross sections to the level of fluorescence cross sections and allow Raman spectroscopy of single molecules. Single molecule Raman spectra can be measured with nonresonant NIR excitation,[31] as well as with resonant excitation exploiting molecular resonance Raman enhancement in addition to SERS.[32] SERS provides a method to detect and identify a single molecule without requiring any label because it is based on the intrinsic surface-enhanced Raman scattering of the molecule. Moreover, it provides structural chemical information and thus the capability to image chemical and physical processes at the single-molecule level without ensemble-averaging effects.

Nonlinear Coherent Raman Spectroscopy. In addition to imaging based on single-photon excited or linear Raman scattering, vibrational images can also be generated using nonlinear coherent Raman spectroscopies. The most prominent nonlinear Raman process for imaging is coherent anti-Stokes Raman scattering (CARS), where molecular vibrations are probed by two incident laser beams, a pump beam at a frequency of ω_1 and a Stokes beam at a lower frequency of ω_2. When the difference between these two frequencies, $\omega_1 - \omega_2$, matches the frequency of a particular molecular vibration, a strong CARS signal is generated at a new frequency, $\omega_3 = 2\,\omega_1 - \omega_2$, higher than both ω_1 and ω_2. This CARS signal depends quadratically on the pump beam intensity and linearly on the Stokes beam intensity and therefore requires picosecond or femtosecond laser pulse trains with high peak powers but only moderate average power for excitation (~ 0.1 mW).[33] A CARS spectrum of a sample, often similar to the spontaneous Raman spectrum, can be obtained by tuning the frequency of the Stokes beam and using a broadband Stokes beam.

Like spontaneous Raman microscopy, CARS microscopy does not rely on natural or artificial fluorescent labels, thereby avoiding issues of toxicity and artifacts associated with staining and photobleaching of fluorophores. Instead, it depends on a chemical contrast intrinsic to the samples. However, CARS microscopy offers two distinct advantages over conventional Raman microscopy:

1. It is orders of magnitude more sensitive than spontaneous Raman microscopy due to the excitation of coherent molecular vibration in the sample. Therefore, CARS microscopy permits fast vibrational imaging at moderate to average excitation powers (i.e., up to ~ 10 mW average power) tolerable by most biological samples. It was found that the peak powers of picosecond laser pulses used for CARS microscopy create minimal nonlinear (multiphoton) damage. Overall, the radiation damage is significantly less for CARS than for spontaneous Raman,

especially when one is interested in following a dynamic process with short data collection time.

2. It has three-dimensional sectioning capability because the nonlinear CARS signal is generated only at the laser focus where laser intensities are highest. Three-dimensional images can be reconstructed by raster scanning the sample layer by layer. This is particularly useful for imaging thick tissues or cell structures.

More Recent Promising Developments in Vibrational Imaging

CARS Microscopy. Although the first CARS microscope was reported in 1982,[34] it was not until the development of a new detection scheme in 1998 that high-quality, three-dimensional images of biological samples became possible.[35] Since then there has been a continuous evolution of the detection schemes and laser sources for CARS microscopy, and the sensitivity has been significantly improved.[36] Many applications in biology and medicine are emerging. For examples, CARS microscopy is used to monitor lipid metabolism in living cells in real time (Figure 3.1) and to image live skin tissue at the video rate.[37]

Raman Imaging of Single Molecules. Single-molecule Raman spectroscopy requires extremely high SERS enhancement factors of at least 12-14 orders of magnitude. The origin of such a level of SERS enhancement is still under debate, but it can be understood as a superposition of an extremely strong electromagnetic field enhancement at factors of about 10^{12} associated with local optical fields and a so-called chemical enhancement effect on the order of ten- to a hundredfold. Due to the mainly electromagnetic origin of the enhancement, it should be possible to achieve a strong SERS effect for each molecule. However, there is still a molecular selectivity of the effect that cannot yet be explained.

A limitation of SERS spectroscopy is that target molecules have to be in the close vicinity of so-called SERS-active substrates such as nanometer-sized silver or gold structures. On the other hand, performing spectroscopy in the local optical fields of the nanostructures provides exciting capabilities for achieving nanoscale resolution in imaging based on Raman contrast. Highly enhanced optical fields are confined at the small probe volume near a metal tip, which is mounted on a cantilever and scanned across a sample surface by atomic force microscopy (AFM). In this tip-enhanced Raman spectroscopy (TERS), the signal enhancement factor is about three orders of magnitude, which is relatively small compared to other SERS experiments. However, the technique allows imaging of single carbon nanotubes with a 25 nm spatial resolution[38] and shows the promise of high-sensitivity Raman microscopy beyond the diffraction limit. The development of new SERS-active nanostructures tailored and optimized for high sensitivity and resolution by nanofabrication techniques is a key goal for future developments of SERS-related chemical imaging.

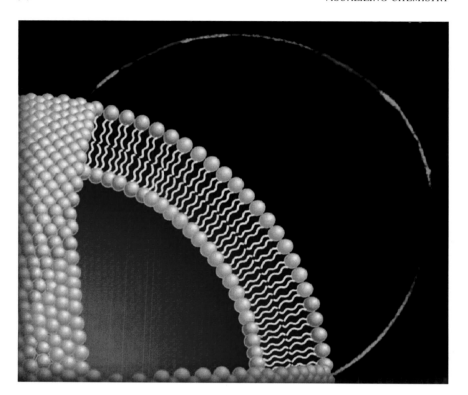

FIGURE 3.1 Image (background) of lipid domains in a single giant unilamellar vesicle (GUV) visualized with CARS microscopy. (Radius of the vesicle = 14 μm.) The high sensitivity of CARS microscopy allows visualization of the GUV's single lipid bilayer when tuning into the frequency of the symmetric stretching vibrational mode of CH_2, which is abundant in lipid molecules. The GUV under investigation has two lipid components, one of which is deuterated. Phase segregation can be clearly observed, as is evidenced by the CARS image of the deuterated lipid (red) overlaid for clarity with the image of the undeuterated lipid (green). The illustration to the left in front indicates the membrane domains (not drawn to scale). For details, see Potma, E.O., and X.S. Xie. 2005. Direct visualization of lipid phase segregation in single lipid bilayers with coherent anti-Stokes Raman scattering microscopy. *Chem. Phys. Chem.* 6:77.
SOURCE: Courtesy of Eric Potma, Harvard University.

New Labels Based on SERS. Design labels with chemical specificity are crucial to imaging.[39] Fluorescence dyes or quantum dots are very common labels, but labels based on SERS signatures for characterizing DNA fragments and proteins have also resulted in high spectral specificity, multiplex capabilities, and photostability.[40] Recently suggested SERS labels made from gold nanoparticles and an attached reporter molecule can provide interesting alternatives to fluorescence tags also for imaging. Figure 3.2 demonstrates simultaneous imaging of the indocyanine green (ICG) gold label based on the SERS signal of ICG along with chemical characterization of the environment of the label by surface-enhanced Raman spectra of the cell components in the vicinity of the gold nanoparticles.[41] The large effective scattering cross section in SERS allows application of very low laser powers (<4 mW) and very short data acquisition times of 1 second or less per spectrum.

Infrared Fourier Transform Microscopy. Infrared Fourier transform (FTIR) microscopy with scanning options allows "chemical mapping" with lateral resolutions of 20 to 60 microns when classical globar light sources are used for broadband illumination. Major advances in recent years in imaging detector technology and step-scan methods have continued to increase the number of applications of IR imaging in materials and biological research.[42] Synchrotron light is a nearly perfect light source for IR spectroscopy because it combines very high brightness and a broad energy range.[43] This results in a considerable improvement in lateral resolutions for synchrotron light, where aperture settings smaller than the wavelength of light can be used and diffraction controls the lateral resolution. For typical IR absorption lines, this means nearly one order of magnitude improvement in resolution compared to a classical light source. This allows the examination of very small dimension structures such as misfolded proteins at very high resolution. For example, the prion protein (PrP) aggregates in scrapie consist of β-sheet structure and are similar to Alzheimer's neuritic plaques; thus, they should be detectable by IR microscopy. However, compared to Alzheimer's disease, PrP aggregates are very small and/or microdisperse in most prion strains. High resolution of these aggregates can be achieved using synchrotron light to monitor this misfolded protein.[44]

Terahertz Imaging. Recent developments of new light sources, particularly free-electron lasers, have led to a rapidly growing interest in using the terahertz range (3-300 cm^{-1}) for imaging.[45] There is considerable evidence that this energy range gives important information on modes related to hydrogen bonds and other weak interactions and can be used for imaging and discrimination among different materials. Moreover, resonances in the terahertz range detected in large biomolecules such as proteins and DNA polymers can provide unique information about the structure of these molecules complementary to that provided by vibrations in the IR frequency ranges and X-ray crystallography. Terahertz transmission spectroscopy of proteins has demonstrated the sensitivity of the technique for monitoring folding-unfolding processes, particularly in a realistic aqueous environment.[46]

At present, free-electron lasers are excellent light sources for basic studies on imaging in the terahertz range. The further development of this technique as a

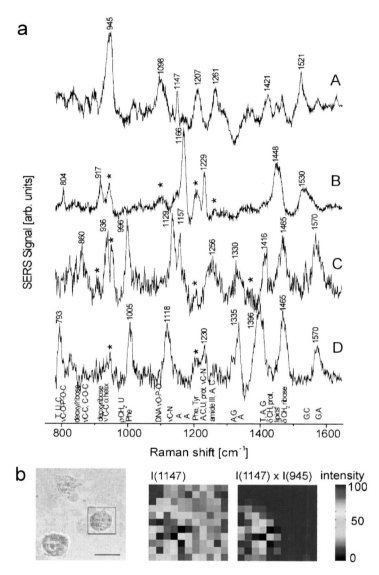

FIGURE 3.2 A hybrid SERS label made from the Food and Drug Administration-approved dye indocyanine green (ICG) on gold nanoparticles and the application of this label inside living cells. (a) Examples of SERS spectra measured from single living cells incubated with the ICG-gold hybrid label at 830 nm excitation. Assignments of major

continued

powerful tool for imaging will also depend on the development of convenient new terahertz light sources.

Future Methodological Developments in Vibrational Imaging

Advances in vibrational imaging techniques are possible in a number of areas. The following is a brief list of the most promising areas of research:

- sensitive vibrational imaging at the single-molecule level, taking advantage of enhanced and confined local optical fields of tailored nanostructures, especially in combination with scanning probe techniques;
- exploration of nonlinear Raman scattering techniques, such as CARS or sum frequency microscopy, in order to enhance vibrational sensitivity in reduced probe volume;
- use of new light sources for IR and Raman imaging that provide higher brightness and tunability over wide wavelength ranges;
- extension of IR imaging to the terahertz range for probing complex macromolecule dynamics and structures and utilizing specific low-frequency modes for high-contrast imaging.

Fluorescence Imaging

Since it began in the seventeenth century, optical microscopy has evolved in capability as a ubiquitous tool of chemistry, biology, materials science, and engineering. From the early days, wide-field microscopy, in which a magnified image of the plane of focus is viewed with visible light through an eyepiece and recorded on film, has been an essential research tool. In particular, the early growth of cell biology, microbiology, and their associated biological chemistry depended on optical microscopy. The inherent physical limits of resolution have frequently

---

FIGURE 3.2 continued

bands are given below spectrum D. ICG bands are marked with an asterisk. (b) Spectral imaging of a SERS label in a living cell based on the SERS spectrum of ICG consisting of several narrow lines. For imaging the label, this offers the advantage that spectral correlation methods can be used to enhance the contrast between the label and the cellular background. A photomicrograph of the cell, indicating the mapped area, is shown for comparison. Scale bar: 20 microns.
SOURCE: Reprinted with permission from Kneipp, J., H. Kneipp, W.L. Rice, and K. Kneipp. 2005. Optical probes for biological applications based on surface-enhanced Raman scattering from indocyanine green on gold nanoparticles. *Anal. Chem.* 77:2381-2385. Copyright 2005 American Chemical Society.

been limiting factors, leading to continuing developments to surpass resolution limits in the focal plane and avoid the out-of-focus background. The contrast in early imaging depended on a variation of refractive index, anisotropic polarizability, light absorption, or scattering. Innumerable advances in capability have followed and are continuing to present major opportunities for advances in chemical imaging that can meet the challenges to solve ever more difficult problems.

Fluorescence Techniques

Fluorescence Microscopy. Unlike NMR spectroscopy and vibrational spectroscopy, electronic spectroscopy involves interactions with electromagnetic waves in the near-infrared, visible, and ultraviolet (UV) spectral regions. While electronic spectroscopy is less enlightening about structural information than NMR and vibrational spectroscopy, the shorter wavelengths involved allow higher spatial resolution for imaging, and its stronger signal yields superb sensitivity. Fluorescence detection, with its background-free measurement, is especially sensitive and makes single fluorescent molecules detectable.

Fluorescence has provided the power and diversity for chemical markers, now frequently designed to bind to particular targets or to be genetically expressed in biological systems. The recent discoveries of green fluorescent proteins (GFPs) and red fluorescent proteins (RFPs) have enabled genetic labeling of particular targets in living systems. This is performed by incorporation of a fluorescent protein gene—from a selection now including many available colors[47] and with chemical properties such as pH[48] or calcium sensitivity[49]—for chemical imaging of physiological functions in vivo.

The continuing development of available labels remains one of the most promising avenues for advances in chemical imaging in both biological and soft materials applications. Figure 3.3 illustrates the GFP-RFP varieties available today. However, they are not optimal because they require slow oxidative activation within the cells in which they are expressed and are therefore not sufficient indicators for many dynamic measurements of gene expression. Current research aims to escape this kind of problem by the development of unnatural amino acids that can be expressed as intracellular markers. This difficult area requires sustained and concerted support.

Detecting and tracking individual macromolecules and their nanoscopic tracking in living cell membranes and tissues provide a powerful approach to understanding the biochemical dynamics of life processes. This approach requires bright fluorescent markers and accurate optical imaging to provide nanoscopic spatial resolution over many orders of time ranging from microseconds to several minutes. However, detection is limited by the number of fluorescence photons that can be captured from each marker before it photobleaches. About 10^3 detected photons are needed at each point in time for a precise location of several nanometers of sparse, and therefore resolvable, markers in an optimized microscope. The

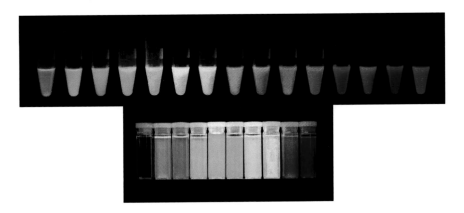

FIGURE 3.3
Top: Various colors of fluorescent proteins now available. These proteins can be expressed in almost any cell and have provided in vivo spectroscopic markers for following the production, degradation, localization, and movement of many different proteins.
SOURCE: Reprinted by permission of Federation of the European Biochemical Societies from Tsien, R.Y. 2005. Building and breeding molecules to spy on cells and tumors. *FEBS Letters* 579:927-932.
Bottom: The broad color spectrum of quantum dots that is now available. These nanoscale particles can be functionalized and attached to a variety of different chemical species for tagging purposes. The great advantages of quantum dots are that they can all be excited at the same wavelength and are very resistant to photobleaching.
SOURCE: Courtesy of Shuming Nie, Emory University.

central position of a marker is obtained by computation of the centroid of the diffraction-limited microscopic image, which is about 450 nm in diameter for a point source.[50] The uncertainty of the position measurement improves roughly as the reciprocal square root of the number of photons detected. Thus, analysis of molecular patterns, mobility, and interactions—important in biological and materials science research—depends on the development of brighter, more durable, chemically specific markers of nanoscopic size in order to label target molecules and follow the time course of their trajectories with nanoscopic precision.

Semiconducting nanocrystals, usually CdSe-ZnS crystals a few nanometers in diameter, called "quantum dots," can be useful for in vivo imaging of biochemical dynamics but still suffer from three limitations.[51] The most severe problem is that quantum dots blink at probability distributions that lead to loss of continuity in keeping track of individual molecules, limiting the ability to measure the dynamics and mechanisms of biophysical chemistry in vivo and ex vivo. Some significant fraction of fabricated quantum dots appear to be totally dark, reducing average quantum yield.[52] Another issue arises from protecting quantum dots from the aqueous biological environment. Various chemical coatings of these hydrophobic particles have been tried, and some thin coatings are temporarily effective. However, the only reliable results to date depend on multiple amphiphilic coatings that increase hydrodynamic diameters to >30 nm, which is too large for many biochemical applications.

To avoid the blinking problem with nonblinking markers and the potential toxicity of semiconducting quantum dots, it is possible to aggregate about 20 to 30 organic dye molecules in protective environments by innovative chemistry. The first success was achieved by labeling low-density lipoprotein (LDL) particles with about 30 carbocyanine dye molecules. These particles can be brightly labeled and photobleach slowly, but they are not durable. A more robust alternative is the sequestration of 20 or so organic dye molecules into a silica shell about 30 nm in diameter using established emulsion techniques. Properly bound to the silica, fluorophores such as rhodamine are protected from photobleaching and interactive quenching, providing a marker about 20 times as bright as present quantum dots but unfortunately also still too large for many applications.[53] However, these two examples demonstrate the potential of utilizing innovative chemistry in the development of more effective chemical imaging tools.

Fluorescence Correlation Spectroscopy and Fluorescence Burst Analysis. Several nanoscopic chemical imaging approaches work very well for measurements of chemical kinetics, interactions, and mobility in solution. Fluorescence correlation spectroscopy (FCS) measures the temporal fluctuations of fluorescent markers as molecules diffuse or flow in solution through a femtoliter focal volume.[54] Their average diffusive dwell times reveal their diffusion coefficients, and additional faster fluctuations can reveal chemical reactions and their kinetics if the reaction provides fluorescence modulation. Cross-correlation of the fluorescence of two distinguishable fluorophore types can very effectively reveal chemical binding kinetics and equilibria at nanomolar concentrations.

These methods work best at nanomolar chemical concentrations so that the focal volume contains typically 1 to 100 molecules on average. Because the method is so sensitive, it is susceptible to perturbation by background fluorescence and instrumentation fluctuations. These problems have become quite tractable during the last decade, such that FCS now supports more than 100 publications per year. A current challenging application is analysis of protein folding kinetics, protein structure fluctuations, and ultrafast chemical kinetics by new methods yet to be published.

Fluorescence burst analysis, a variation of FCS procedures that has an optimum configuration for simple presentation, uses a uniform nanoscopic flow channel with an optically perfect ceiling, uniform cross section, and periodic electrodes that can now be constructed by careful electron lithography techniques.[55] By application of controlled electric fields, uniform plug flow of solution through the channel is achieved by controlled electrophoresis. This avoids the parabolic flow velocity profile of pressure-driven flow. The channel cross section is uniformly illuminated by sufficiently large laser beam flows to provide identical illumination pathways for molecules flowing anywhere in the cross section. Thus, all molecules of a given brightness, such as a particular length of DNA labeled with an intercalating dye, yield the same fluorescence burst size, thereby providing a burst size parameter that characterizes the DNA double helix length to approximately ±5 percent over at least three orders of magnitude. Clearly, future applications of this "flow imaging" burst analysis scheme offer potential for development of analytical techniques in medicine such as the elusive counting of the concentration of beta-amyloid clusters and their sizes in cerebral-spinal fluid.

Single-Molecule Fluorescence Spectroscopy and Imaging.

Current Technology. In the past decade, rapid developments have made it possible to detect, identify, track, and manipulate single molecules on surfaces, in solutions, and even inside living cells. The ability of single-molecule experiments to avoid ensemble averaging and to capture transient intermediates make them particularly powerful in elucidating mechanisms of molecular machines in biological systems: how they work in real time, how they work individually, how they work together, and how they work inside live cells. New knowledge from single-molecule experiments continues to generate novel insights in a variety of scientific fields.

Single-molecule fluorescence detection in an ambient environment is achieved in part through reduction of the probe volume in order to suppress the background signal.[56] This is accomplished by a confocal or total internal reflection microscope and as well as by the high sensitivity of the detectors. Aside from the fluorescence intensity, optical properties such as polarization, fluorescence lifetime, and excitation and emission spectra have been used as contrast mechanisms for acquiring images to follow the temporal behavior of certain molecules. In particular, fluorescence resonance energy transfer has been widely used as a dynamic variable, dubbed a "molecular ruler," to measure intermolecular distances between two fluorophores. These advances allow one literally to record movies of molecular motions and biochemical reactions.

One of the exciting areas of research with fluorescence microscopy is the study of dynamic behaviors of individual enzyme molecules. Conventional measurements of chemical kinetics rely on determining concentration changes following a perturbation (such as a temperature jump or rapid mixing of reactants). On a single-molecule basis, a chemical reaction, if it occurs, takes place on the subpicosecond time scale. However, the "waiting time" prior to such an action during which the molecule acquires energy to reach the transition state via thermal activation is usually long and stochastic. Stochastic events of chemical changes can be monitored and the histogram of waiting times can be measured for a single enzyme molecule undergoing repetitive reactions. The advantages of single-molecule studies of biochemical reactions include the following: (1) to measure the distributions and fluctuation of enzymatic activities; (2) to unravel reaction mechanisms; and (3) to observe in real time the transient intermediates that are otherwise difficult to capture in conventional experiments due to their low steady-state concentrations.

For example, enzymatic turnover of flavin enzyme molecules was monitored in real time by viewing fluorescence from an active site of the enzyme.[57] Cholesterol oxidase, a 53-kilodalton flavoprotein, catalyzes the oxidation of cholesterol by oxygen with the enzymatic cycle shown in Figure 3.4. The active site of the enzyme, flavin adenine dinucleotide (FAD), is naturally fluorescent in its oxidized form but not in its reduced form. Confined in agarose gel containing 99 percent water, the enzyme molecules are immobilized at the laser focus. On the other hand, the small substrate molecules are essentially free to diffuse within the gel. The single FAD emission exhibits on-off behavior, with each on-off cycle corresponding to an enzymatic turnover. This experiment demonstrated that an enzyme molecule is a dynamic entity with a fluctuating catalytic rate constant, a phenomenon that was hidden in ensemble studies.

The conformational dynamics of enzymes is intimately related to enzymatic activity and can now be probed at the single-molecule level. Fluorescence resonant energy transfer (FRET) is used widely in biochemical and biophysical studies of conformational motions.[58] The efficiency of FRET between a donor and an acceptor pair is $E = 1/(1 + (R/R_0)^6)$, where R is the distance between the pair and R_0 is the Forster radius, which is dependent on the spectral overlap between the donor emission and acceptor absorption spectra and the relative orientations of the donor and acceptor dipoles. FRET between a single donor and acceptor pair within a single biomolecule has been used to probe conformational dynamics.[59] For example, a small RNA enzyme called the hairpin ribozyme has been studied.[60] The ribozyme's two domains were labeled with a FRET pair, and the FRET time traces showed striking heterogeneity in docking and undocking kinetics, suggesting the presence of a large number of stable conformational states under functional conditions.

For experiments with fluorescent substrates, substrate concentrations must be kept low to avoid a strong fluorescent background. At millimolar substrate concentrations where many enzymatic reactions occur, conventional FCS would not work at the usual femtoliter focal volumes. To escape this limitation, it has been possible to provide attoliter focal volumes in electron lithographically formed zero-mode waveguides. For example, these structures have allowed the tracking of the formation of the complementary DNA sequence for a template sequence by high-processivity function of a single DNA polymerase.[61] Essentially, this geometry provides the opportunity for virtual single-molecule enzyme kinetics at appropriate fluorescent substrate concentrations wherever a non-interfering signal can be created.

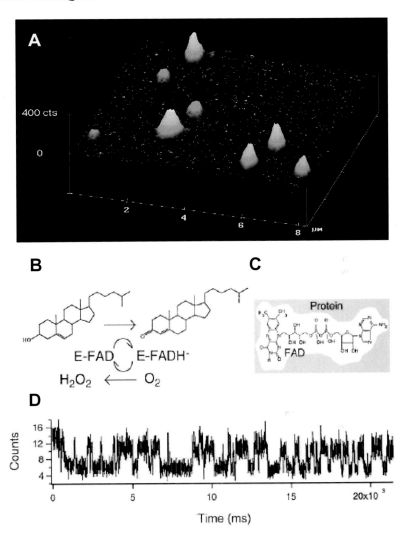

FIGURE 3.4 (A) Fluorescence image of single cholesterol oxidase (COx) molecules immobilized in a thin film of agarose gel of 99 percent buffer solution. (B) Enzymatic cycle of COx that catalyzes the oxidation of cholesterol by molecular oxygen. The enzyme's naturally fluorescent FAD active site is first reduced by a cholesterol substrate molecule, generating a nonfluorescent FADH⁻, which is then oxidized by molecular oxygen. (C) FAD, the fluorophore and active site of COx. (D) A portion of the intensity trajectory of an individual COx molecule undergoing enzymatic reactions in real time. Each on-off cycle of emission corresponds to an enzymatic turnover.

SOURCE: Reprinted with permission from Lu, H.P., L. Xun, and X.S. Xie. 1998. Single-molecule enzymatic dynamics. *Science* 282:1877-1882. Copyright 1998 AAAS.

Parallel with and complementary to single-molecule studies by optical means has been tremendous work on mechanical manipulation of single molecules accomplished through the use of either optical tweezers[62] or magnetic tweezers.[63] These techniques offer the possibility of actively controlling the behavior, or even chemical reactions, of single molecules and have yielded much new knowledge about the mechanisms of enzymatic machineries such as molecular motors[64] and nucleic acid enzymes.[65]

Cutting-edge Technology. Integrating chemical and biological labels with advanced microscopes and detectors is the focus of many research activities. Other contrast mechanisms of single-molecule imaging, in addition to fluorescence, are also being pursued.

The spatial resolution of fluorescence microscopy has been limited to about half the wavelength of light due to the diffraction limit associated with the wave nature of light. However, if one has a single isolated molecule with bright fluorescence, the accuracy of determining the center position of its diffraction-limited image can be as high as 1 nm. In this way, nanometer movements of a molecular motor can be followed in real time.[66]

At high concentration, when molecules are no longer isolated in space, a conventional optical microscope is unable to resolve them within the diffraction limit. Efforts have been made to circumvent the diffraction limit by engineering the point spread function using nonlinear optical techniques. Spatial resolution of 20 nm in a cell has been demonstrated without using a proximal probe.[67]

Future Technology. Recent innovations of single-molecule fluorescence imaging and dynamical studies have lead to unprecedented sensitivity, molecular specificity, time resolving power, and spatial resolution. In the future, the challenges and opportunities in optical imaging will lie with biology. Single-molecule sensitivity for three-dimensional imaging in a living cell with specific and noninvasive labeling of a macromolecule of interest, along with millisecond time resolution and nanometer spatial resolution, will provide answers to many biological questions. The integration of these elements will come with time, and a motion picture of a living cell should prove possible with continuing developments.

Laser Scanning Microscopies

Confocal Microscopy

The availability of laser scanning confocal fluorescence microscopy,[68] first commercially offered in the 1980s, enabled a major advance in chemical microscopy imaging; convenient summaries are available in the *Handbook of Biological Confocal Microscopy.*[69] Confocal fluorescence microscopy works by focusing continuous-wave laser illumination to a diffraction-limited spot in a focal plane within the specimen and collecting the excited fluorescence through a confocal optical aperture that excludes most of the out-of-focus fluorescence background. Images are formed by scanning the laser beam in a video raster and recording the photomultiplier-detected fluorescence in a computer array. Lateral resolutions are comparable with wide-field microscopy, and effective axial resolution is enhanced by exclusion of out-of-focus background. The technique allows imaging to depths of about 50 microns in soft biological tissue but is limited by the background-scattered fluorescence able to pass through the confocal aperture. This technology is used widely today in cell biology and soft-matter materials using fluorescent markers for chemical imaging.

Multiphoton Microscopy

Laser scanning fluorescence microscopy entered a new generation in 1990 when the nonlinear optical physics of two-photon molecular excitation (first analyzed by Maria Goeppert-Meyer in 1931 but not demonstrated until sufficiently bright lasers were created in the 1960s) was finally formulated for useful multiphoton laser scanning fluorescence microscopy.[70] Multiphoton excitation of fluorescence provides several critical advantages over wide-field and confocal microscopy. Because multiphoton excitation of a molecule requires that it "simultaneously" absorb two or more excitation photons, fluorescence excitation is typically limited to the focal volume where concentration of the laser power provides sufficient photon flux density. Since the two-photon excitation rate depends on the square of the illumination intensity, the out-of-focus background excitation falls as the reciprocal fourth power of distance above and below the focal volume, thus generating negligible out-of-focus fluorescence along the out-of-focus beam path. Photodamage is also negligible since the long laser wavelengths needed for multiphoton excitation (nearly invisible infrared photons) are not significantly absorbed by tissue. This nonlinear microscopy can productively image fluorescence signals to depths in living tissue of approximately 500 microns (about the thickness of human skin).

Multiphoton microscopy (MPM) utilization has grown rapidly and continuously since then, with more than 200 refereed publications per year citing the use of MPM or two-photon microscopy. Commercial sources for MPM instruments did not become available until several years later, but adequate titanium sapphire (Ti:sapphire) 100-femtosecond lasers were (and are) available, albeit at exorbitant costs. Many MPM instruments were and still are assembled by the scientists using them, a point that may become relevant in future specialized chemical imaging opportunities. Convenient laboratory instruments for MPM imaging are now available from Zeiss Microscopy.

The earliest, fastest-growing, and possibly most productive area of application of MPM is in the imaging of neuronal functions in ex vivo functional brain slices and protracted imaging of function in intact brains of living animals over extended times as the neural circuits develop.[71] The most popular chemical applications of MPM have been based on fluorescent molecular indicators of calcium ion activity, a ubiquitous intracellular signal, and of membrane potential. Recently, GFP gene labels of specific receptors and ion channels and fluorescent labels of protein active sites involved in the molecular mechanisms of biological functions have provided additional powerful research tools.[72] The development of three-photon infrared excitation of the intrinsic UV excitable indoleamines, serotonin and melatonin allows research to be conducted on secretory kinetics and mechanisms for neuromodulator release in cell cultures and in living tissue.[73] This capability has yet to be fully realized for in vivo or ex vivo studies of the secretion of these important neuromodulator molecules in brain.

Deeper Multiphoton Fluorescence Imaging in Living Tissue Through GRIN Lenses. It is possible to translate the focal volume of MPM imaging by up to 0.5 cm distances using gradient refractive index (GRIN) lenses. These lenses consist of small-diameter rods of exotic, rare-earth-containing glasses of graded concentrations that provide a radial gradient of refractive index, thereby acting as a lens with flat ends. Multiphoton images at depths up to about 0.5 cm in the brains of living mice have been obtained with access to the intact mouse hippocampus and negligible tissue damage en route.[74] The longer GRIN rods for focusing transfer are a few millimeters in diameter and are capped by a short, higher-numerical-aperture objective lens. It is possible to miniaturize these devices by using extensions of current techniques and delivering the femtosecond laser pulses with suitable fiber optics using vibrational scanning.[75] This technique appears to have great promise for deeper in vivo chemical imaging.

Intrinsic Biological Fluorescence and Potential Applications in Medicine. The most recent and potentially most important advances in MPM are based on chemical imaging of the intrinsic fluorescence of crucial molecular species. One, in particular, images the long-known fluorescence of nicotinamide adenine dinucleotide (NADH) to measure the metabolic pattern in brain, recognizing oxidative exhaustion of neurons caused by their electrical signaling activity and the slower contribution to restoration of their metabolism by astrocytic glycolysis.[76] Other possible chemical signals that supposedly couple astrocytes with neurons in brain function were entertainingly but significantly summarized in *Scientific American* in April 2004.[77] An interesting challenge of metabolic imaging is the chemistry of NADH-NAD(P)H and their binding to protein cofactors in the mitochondria and cytoplasm, which modulates the fluorescence quantum yield and confounds the accuracy of quantitative measurements of metabolic state.[78] In addition, this topic will be a chemical imaging challenge that must be solved in coming years, since MPM imaging of brain metabolism in living animals, including neurodegenerative disease models, now appears to be approaching feasibility.

Second Harmonic Generation (SHG) Imaging

Another nonlinear optical technique, known as SHG, can be imaged with bright-pulsed laser illumination of optically noncentrosymmetric materials with nonlinear (intensity-dependent) dielectric properties. Certain amphiphilic or hydrophobic electropolarizable dye molecules that are lipid soluble can be aligned in parallel by the electric fields commonly present across cell membranes. These fields routinely reach up to 250,000 volts per centimeter, a signal of great significance for controlling cellular behavior in neurons. The demonstrated effectiveness of SHG fast imaging of neuronal signals motivates efforts to develop further improvements of the noncentrosymmetric electric field-sensitive indicator molecules. There have been sustained international efforts to develop membrane potential sensitive fluorescent molecules, but SHG electric field indicators are a relatively fertile photochemical challenge.

SHG has been found to provide a selective marker for imaging neuronal axons through SHG generation by the parallel-oriented bundles of their microtubules, which provide the selective tracks for movement of cargo in vesicles to and from synapses by the molecular motors dynein and kinesin.[79] The parallel polarization of microtubules in axons and their random orientation in neuronal dendrites had previously been detectable only by tedious multistage electron microscopy. This new capability may be useful in diagnosing the effects of aggregation of the microtubule-associated tau protein (imageable by its MPM intrinsic fluorescence) that induces neurofilamentary tangles in Alzheimer's disease.

SHG imaging of collagen structures has been very effectively achieved and is now rather well understood.[80] Combining SHG and MPM fluorescence appears feasible for creation of a simple optical label of collagen chemical assembly structure type and anomalies in orthopedic surgery, discussed further below.

Other Multiphoton Coherent Optical Microscopy

In a manner similar to SHG imaging, third harmonic generation (THG) imaging has been demonstrated.[81] CARS microscopy (discussed earlier) is another form of MPM, providing chemical information via vibrational spectroscopy. As in two-photon fluorescence microscopy, SHG, THG, and CARS techniques have small probe volumes and offer three-dimensional resolution. However, unlike two-photon fluorescence, SHG, THG, and CARS signals (like those of NMR) are coherent. Therefore, the contrast interpretations in SHG, THG, and CARS microscopy are more complicated than two-photon fluorescence microscopy. In recent years, these techniques have been studied experimentally and theoretically in great detail.[82] The spectral specificity and high sensitivity of CARS microscopy are particularly attractive for chemical imaging of living cells.

Ultrafast Spectroscopy and Imaging

The advent in recent years of pulsed light sources that generate femtosecond pulsed lasers that are easily operated, are computer controlled, and can be integrated into imaging systems has provided new capabilities in imaging spectroscopy in two respects. First, and thus far most important, the high peak power that such pulses provide has made it possible to perform nonlinear imaging spectroscopy routinely. The most popular version of this, and the only commercially available approach, is multiphoton fluorescence microscopy. Other nonlinear imaging techniques such as sum-difference spectroscopy of surface molecules have been developed as well but are not yet commonplace. Another current use of ultrafast pulses in imaging is the measurement of excited state dynamics. In particular, fluorescent lifetime imaging has become popular in the biological community because the fluorescent lifetime is sensitive to the environment but does not depend on the concentration of the label. Finally, ultrafast pulses are being used for tomography of deep tissue, primarily in the realm of analyzing light transmitted through a scattering sample.

Current Capabilities

Fluorescence Lifetime Imaging (FLIM). Commercial instruments are now available for FLIM of surfaces or three-dimensional objects in conjunction with either multiphoton or confocal microscopy. At each point in the image, an excited state dynamics trace is measured, typically on the few hundred picosecond to nanosecond time scale. This type of imaging is particularly useful in conjunction with FRET systems. When the donor and acceptor become close to one another, the fluorescence lifetime of the donor decreases as the energy transfer rate becomes comparable to, or faster than, the inherent donor excited state lifetime (Figure 3.5).[83] Other labels have lifetimes that are particularly sensitive to specific environments. For example, DNA intercalating dyes such as thiazole orange have very short lifetimes in solution, but much longer and somewhat variable lifetimes when intercalated into DNA. This allows one to determine not only whether the signal comes from DNA but also the state of the DNA (e.g., base content and structure).[84] Finally, almost all organic dyes have inherent environmentally sensitive fluorescent lifetimes and can be used to report on conformational changes in proteins or other molecules in an imaged fashion.[85] FLIM provides another dimension to distinguish between the signal from the label under study and background fluorescence. In the case of biological imaging, autofluorescence is usually relatively short-lived compared to most of the organic dyes employed. In addition, FLIM provides increased resolution between labels that have similar spectra but different lifetimes, allowing the ability to separate many fluorophores from one another in a sample simultaneously using a two-dimensional (time versus spectrum) approach.

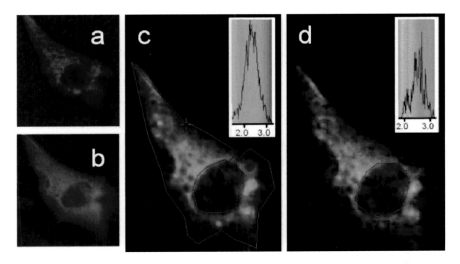

FIGURE 3.5 Example of the use of FLIM to investigate molecular interactions in cells. Two-photon FLIM was performed on GFP-tagged protein kinase C (GFP-PKC) coexpressed with DsRed-tagged caveolin (DsRed-cav) in Chinese hamster ovary (CHO) cells. Coexpression of the GFP-PKC with DsRed-cav does not affect the lifetime of the GFP, showing that in the unstimulated state, PKC is not associated with caveolin. Epifluorescence images for excitation of (a) DsRed and (b) GFP along with DsRed show the PKC and caveolin codistributed in the cytosol. (c) Fluorescence lifetime images with the analysis area enclosed by the red line (cytosol) or nucleus both essentially show a lifetime centered around ~2.2 nanoseconds. Spatial scale not specific for image in the original figure; however, average CHO cell diameter is approximately 15 μm.
SOURCE: Stubbs, C.D., S.W. Botchway, S.J. Slater, and A.W. Parker. 2005. The use of time-resolved fluorescence imaging in the study of protein kinase C localization in cells. *BMC Cell Biol.* 6:22. Copyright 2005 Stubbs et al; licensee BioMed Central Ltd. This is an Open Access article distributed under the terms of the Creative Commons Attribution License.

Ultrafast Tomography. Another developing imaging technique is tomography using ultrafast near-infrared pulsed lasers.[86] This can be done in several ways. One basic approach involves detecting transmission of ultrafast pulses through tissue. Here, one is looking specifically at light that travels directly through the tissue and therefore is not delayed in its arrival at the detector. This is performed by gated detection methods; in this way, it is possible to generate a two-dimensional projection image that includes information about absorption and scattering density in the tissue. Ultrafast pulses can also be used specifically to detect light scattered from a particular depth in the tissue by timing the round-trip.

Enhanced penetration capabilities of wide-field microscopy[87] have been obtained by elegant application of interferometry in a technique called optical coherence tomography[88] that images optical backscatter contrast in tissue at millimeter depths. This powerful technique is already being applied in medicine, particularly to image the retina within the eye, an especially favorable application for deep imaging since the optical path has little scattering. Interferometry, beginning with early designs of Zernike and Nomarski, has been extended to picometer displacement measurements of the transduction mechanisms of audition and their biochemical modulation.[89] Image resolution enhancement by about twofold of thin biological preparations has been obtained by computational convolutions of stacks of images.[90]

Cutting-edge Technology

Ultrafast spectroscopic approaches are increasingly being applied to imaging in terms of both nonlinear imaging and dynamics. This is driven partially by the improvement in and ease of use of ultrafast laser systems and detectors. It is not possible to review all of the methods currently under development, but several examples are given below.

Spectral Lifetime Imaging. Because of improvements in detector sensitivity, multidimensionality, and speed, a number of different ultrafast methods have been converted into imaging methods. A particularly interesting example is the use of two-dimensional streak cameras in fluorescence imaging. The streak camera allows one to record real-time signals with subpicosecond resolution in one dimension while simultaneously resolving another dimension such as wavelength. This provides an opportunity to record the fluorescence decay time and the spectrum with high resolution at each point in a three-dimensional image using confocal microscopic techniques. This approach, coupled with hyperspectral analysis in both dimensions, has the potential for extraordinary resolution of different fluorescent targets in a complex sample mixture.[91] The addition of a spectral dimension to a typical fluorescence microscope can provide an increase in sensitivity, throughput, and data accuracy.

Reverse Imaging. In imaging, one normally thinks of obtaining information about a system that is spatially and/or temporally resolved. However, similar approaches can also be used to project patterned information into a chemical system. This, for lack of a better phrase, is referred to as "reverse imaging" in this document. The idea is to use the same kinds of methods and machines that we use to record images to also control chemistry or biochemistry in a patterned fashion.

Current Applications

Chemistry. Photolithography has been refined for several decades by the electronics industry. Three-dimensional sculpting of photopolymerizable materials using nonlinear excitation has allowed the production of objects with resolutions below 100 nm. Applications to patterning of chemical systems can be seen in the photoprocessing of DNA chips (see Figure 3.6, for example) by companies such as Affymetrix (Santa Clara, CA). Light is used to build large arrays of DNA oligonucleotides layer by layer in the horizontal plane. Similar approaches are now being used to pattern chemistry using electrochemical methods on electrode arrays (Combimatrix, Seattle WA). However, this technology has been limited largely to polymerization of homogeneous photopolymers and production of DNA. The potential for developing chemical systems that can be modulated in time and space by imaged radiation is huge.

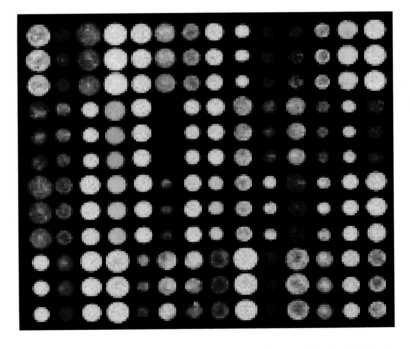

FIGURE 3.6 Example of a DNA array and fluorescent detection. Synthesis of chips such as these is valuable in determining changes in gene expression under various physiological conditions with high throughput.
SOURCE: Photograph courtesy of Peter R. Hoyt, Oklahoma State University Microarray Core Facility, and Mitchel J. Doktycz, Oak Ridge National Laboratory Life Sciences Division.

Biology. In biology, studies have been conducted using imaged killing of cells either for directed evolution or for tissue engineering. Also, laser-based photoablation of cells has been used in fate-mapping studies of development. Subcellular surgeries have been performed in which specific features of a cell have been ablated. However, these are crude approaches since the method used in these procedures is typically photodestruction rather than specific manipulation.

An important technique in current use is image-active photochemistry. This is employed in the use of chemical caged derivatives of neurotransmitters and biological activators such as adenosine 5'-triphosphate (ATP), which can be used as powerful optical tools for photoactivated research pharmacology. Effective chemical cages for one-photon photoactivation are numerous and widely used. The continuing advance in development of these photochemical tools indicates the potential for further chemical advances. Unfortunately, two-photon excitation of these same caged reagents has not worked as well as the best reagents for one-photon excitation, thus excluding the pharmacology envisioned in vivo.[92] These obstacles could be overcome by utilizing new cage designs or three-photon excitation of the UV-activated cages to avoid both parity selection rules and pulse pairs for multiple-stage cage release kinetics. This is a challenge for future fast chemical imaging activated approaches.

Laser tweezers can also be used to manipulate specific parts of cells in ways much more subtle than photodestruction. It has also become possible to design genes in which expression is controlled by light. This opens the door for patterned gene expression at cellular resolution in both two and three dimensions. In the chemical world, the future holds many opportunities for conducting two- and three-dimensional solid-phase synthetic chemistry using light. A fairly large arsenal of photoprotective groups now exists, allowing a wide variety of chemical couplings to be performed. This may revolutionize the search for new drugs, sensors, or materials because very large libraries of molecules can be synthesized rapidly. The advantage over standard combinatorial chemistry is that these are not random combinations of pieces but specific molecular structures, allowing one to combine rational design driven by computation with very high throughput screening and selection during molecular evolution.

Prospectus for Future Optical Microscopy

Which aspects of chemical imaging by optical microscopy will be most important in coming years? What needs to be done to take advantage of its powers? In vivo and ex vivo chemical imaging in the brain and nervous system will continue to be important. Current applications are being extended to chemical diagnostics of biochemical signaling between dendritic spines and presynaptic patches on axons that defines the development of neural circuits during growth and learning of the brain.[93] The principal challenge of neuroscience research in the coming decades will be discovery of the chemical signals governing the development of nervous system organization and the mapping of its consequent geometry and function.

Intrinsic fluorescence of tissue components is imaged efficiently by MPM in living animals and tissues, with results that demonstrate substantial capability for instant recognition of cancer in several organs.[94] Direct comparisons between the incipient tumor MPM images in organs at risk with the corresponding pathologists' images of absorption-stained fixed tissues show that MPM is quite functional for disease diagnosis and study.

Medical applications for use in surgery and diagnosis are already being developed in various collaborations at several institutions and will certainly expand worldwide. Two basic challenges must be met: (1) the need for increasing knowledge of biological photochemistry and (2) the development and engineering of suitable optical physics for devices compatible with the restrictions, demands, and safety of medical and surgical procedures.

The crucial biological photochemistry issue is to understand in greater detail all of the origins of intrinsic tissue fluorescence and its changes with disease. It is clear that NADH (an indicator of metabolic state) and the many flavin compounds that vary greatly in tissue types will be among the most conspicuously imaged indicators of disease. Other fluorescence contributors may include carotinoids and oxidized indoleamines. Imaging and macroscopic mapping of different types of collagens using second harmonic generation also appears promising as a medical tool, especially for orthopedic diagnostics and surgery.

General medical applicability of MPM calls for more compact, rugged, and user-friendly instrumentation. Ideally, designs could be built onto the observation optics currently used in surgery, particularly the endoscopic optics of laparoscopic surgery. The possibility of achieving this adaptability is advanced by (1) the discovery of rugged fiber optics capable of single-mode propagation by the high-power femtosecond laser pulses needed for MPM,[95] and (2) the development of vibrating optical-fiber scanners for illumination scanning (Optiscan, Inc., Australia). Endoscopic MPM microscopy, which effectively provides in situ chemical and structural imaging, appears capable of instantly providing image information equivalent to conventional fixed-tissue pathology imaging.

Fundamental advances in research on complex chemical process are being enabled by new nanoscopic imaging developments that enable measurements of (1) chemical kinetics of protein folding and misfolding (e.g., amyloid aggregate formation in neurodegenerative diseases); (2) fluctuations in single-molecule enzyme kinetics to understand multistep enzymatic reactions; and (3) ultrafast chemical kinetics. These processes and their possibilities for application represent opportunities to be exploited via new developments in nanoscopic and nonlinear optical imaging discovery.

ELECTRON, X-RAY, ION, AND NEUTRON SPECTROSCOPY

This section covers techniques that probe samples with wavelengths much smaller than that of visible light and that provide high-resolution chemical and structural information below surfaces of materials. For example, X-rays are able to penetrate materials so deeply it is possible to determine the identity and local configuration of all the atoms present in a sample. Further attributes and limitations of these techniques are provided below.

Electron Microscopy

Since its invention nearly 100 years ago, electron microscopy (EM) has developed into a remarkably versatile imaging tool. Nearly all major R&D organizations (universities, national laboratories, industrial labs) have EM facilities that can accommodate well-established imaging techniques. Despite its apparent maturity, new techniques continue to be developed that not only push the resolution limits of EM but also expand the range of specimens and environments in which it can be used.

The usefulness of electrons for imaging comes from the fact that an electron wavelength is about 1000 times smaller than that of visible light, providing a much higher-resolution probe. In addition, higher-energy electrons (on the order of hundreds of kiloelectronvolts, or keV) can penetrate materials, thus providing access to imaging below their surfaces. Similarly the use of light, energy-resolved electron detection is able to provide chemical information in parallel to structural imaging. However, significant limitations to its use do exist, such as the need for a vacuum to produce and transmit electrons and electron beam damage to samples. Nevertheless, ingenious development of a variety of EM techniques has had tremendous impact in fields ranging from condensed matter physics to structural biology.

One of the most important forms that EM technology takes is the transmission electron microscope (TEM). The TEM operates much like a slide projector in the sense that electrons of sufficiently high energy (usually in the range of a few hundred kiloelectronvolts) are passed through a thin sample (usually less than a micrometer thick) to a detector where a variety of imaging schemes can be implemented. A number of applications of TEM are described below.

Cryo-Electron Microscopy for Biological Structure Determinations

The use of electron microscopy to image biological systems is a fast-growing area. One advance is the use of rapid freezing to transfer hydrated specimens into an electron microscope, a process known as cryo-electron microscopy (cryo-EM). After such a transfer, new techniques of imaging may be employed to display three-dimensional structures over a length scale ranging from single biomolecules to synapses and neurons.[96] Three-dimensional structure determinations based on cryo-EM have become a standard tool of structural biology in recent years.[97] As in crystallography,[98] the technique of freezing samples in vitreous ice for EM analysis[99] has made it possible to obtain two-dimensional projected images with minimal distortion or artifacts. If there is a plentiful supply of nearly identical frozen particles in random orientations, these projections can be combined to form three-dimensional images. The resolution of these images has been improving rapidly due to improvements in reconstruction techniques.[100] In particular, efforts have centered on (1) the accurate determination of the contrast function that corrects the two-dimensional images for the experimental out-of-focus distance; (2) the accurate determination of the relative orientation of the projected images; and (3) the use of a far greater number of particles. As a result, it is now routine to obtain cryo-EM image reconstructions to have an estimated resolution of 10 Å (sometimes down to 7 Å) with expectations of reconstructions as low as 4 Å.[101]

Recent advances in instrumentation, data collection, and data analysis have resulted in cryo-EM maps of the ribosome with significantly higher resolution, as shown in Figure 3.7. For the first time, molecular signatures can be recognized: these include RNA helices with distinctly visible major and minor grooves, and protein domain structure with clearly defined shapes. The high degree of definition of these features has made it possible to fit known atomic structures with an accuracy of 3 Å.

Scanning Transmission Electron Microscopy

Conventional imaging using TEM has provided a tremendous capability to image atomic arrangements over a large range of length scales. To investigate the chemical nature of materials, however, elemental resolution is required. Developments in the formation and positioning of atomic-sized electron beams have enabled the development of scanning transmission electron microscopy (STEM) in which the highly focused probe beam is rastered across the sample with atomic level control. This has enabled the development of Z-contrast imaging in which the intensity of the formed image is directly related to the atomic number Z. As a result, elementally sensitive imaging can be performed with atomic-level resolution. Pennycook and colleagues demonstrated 0.6 Å resolution using Z-contrast STEM in imaging columns of silicon atoms (Figure 3.8).[102]

Further chemical information can be extracted by the use of an energy analyzer in the STEM. Electron energy loss spectroscopy (EELS) can be performed in conjunction with STEM to provide chemical bonding information. This has been extremely useful for investigating the chemical nature of defect structures such as grain boundaries in ceramics. Klie and Browning have examined the atomic structure, composition, and bonding of grain boundaries in strontium titanate ($SrTiO_3$) and found segregation of oxygen vacancies to the grain boundary

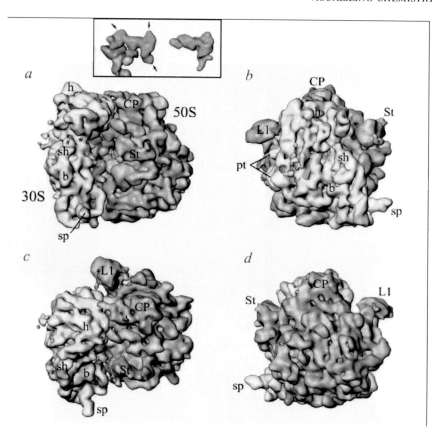

FIGURE 3.7 A cryo-EM map of the *Escherichia coli* ribosome (complexed with fMet-tRNAf Met and mRNA); where fMet = formylmethionine obtained from 73,000 particles at a resolution of 11.5 Å. (a-d) Four views of the map, with the ribosome 30S subunit painted in yellow, the ribosome 50S subunit in blue, helix 44 of 16S RNA in red, and fMet-tRNA at the P site in green. Inset on top juxtaposes the experimental tRNA mass (green, on left) with the appearance of the X-ray structure of tRNA at 11 Å resolution (on right). Arrows mark points at which tRNA contacts the surrounding ribosome mass. Landmarks: h = head and sp = spur of the 30S subunit. CP = central protuberance: L1 = L1 stalk and St = L7/L12 stalk base of the 50S subunit.

SOURCE: Reprinted with permission from Gabashvili, I.S., R.K. Agrawal, C.M. Spahn, R.A. Grassucci, D.I. Svergun, J. Frank, and P. Penczek. 2000. Solution structure of the *E. coli* 70S ribosome at 11.5 Å resolution. *Cell* 100:537-549. Copyright 2000, with permission from Elsevier.

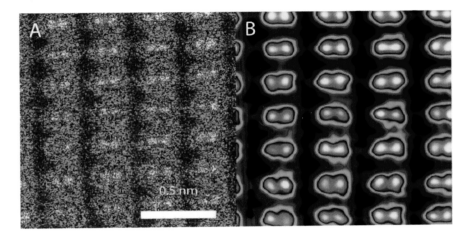

FIGURE 3.8 (A) Annular dark-field images of a silicon crystal using Z-contrast STEM. The image has been low-pass filtered to reduce the noise, and the small effects of image drift during the scan have been unwarped. This direct, subangstrom-resolution image shows dumbbell-shaped rows of atoms with a spacing of 0.78 Å between each pair. Analysis of the power spectrum shows the presence of information down to a record 0.6 Å. The image was obtained with Oak Ridge National Laboratory's aberration-corrected 300 kV Z-contrast scanning transmission electron microscope.

SOURCE: Reprinted with permission from Nellist, P.D., M.F. Chisholm, N. Dellby, O.L. Krivanek, M.F. Murfitt, Z.S. Szilagyi, A.R. Lupini, A. Borisevich, W.H. Sides, Jr., and S.J. Pennycook. 2004. Direct sub-angstrom imaging of a crystal lattice. *Science* 305:1741. Copyright 2004 AAAS.

that is increased at elevated temperatures and is independent of the cation arrangement.[103] These measurements provide direct support for recent experimental and theoretical predictions that nonstoichiometry, and in particular oxygen vacancies, can be responsible for widely observed grain boundary properties.[104]

In Situ Imaging: Solid-Liquid Interfaces and Epitaxial Growth

The chemistry of solid-liquid interfaces lies at the heart of numerous fields of chemistry, chemical engineering, and biochemistry. The ability to image at this interface has proven a tremendous technical challenge. Despite the apparent incompatibility of electrons with liquids, new advances in TEM have been developed that permit direct imaging of a reacting solid-liquid interface, namely, electrochemical deposition. The chemistry of electrodeposition is vital to a variety of technologies ranging from coatings to microelectronics, and high-resolution imaging of this process would greatly improve our understanding of this chemistry. An important technical challenge has been the development of a working electrochemical cell within the vacuum environment of a TEM that would be thin enough for electron transmission. Ross and colleagues have made progress in this area by manufacturing a micron-scale electrochemical cell that allows them to image the electrodeposition of copper in situ at video rates using TEM.[105] The formation, growth, and dissolution of individual, nanometer-scaled clusters were imaged quantitatively, allowing an analysis of the nucleation and growth process for copper plating.

While high-energy, penetrating TEM has proven useful for the in situ imaging of solid-liquid interfaces, lower-energy implementations of electron microscopy are ideal for imaging processes such as thin-film epitaxy. Low-energy electron microscopy (LEEM) is a relatively new technique that takes advantage of the surface sensitivity of electrons in the 0 to 100 eV energy range. Electrons in this range penetrate only a few atomic layers into a specimen before being reflected. The principle of LEEM is directly analogous to that of traditional light microscopy in the sense that low-energy electrons are beamed at the specimen and reflected back, and by using a series of lenses, a real space image is formed. A key aspect of this technique is that video-rate images can be obtained of processes that can occur over a wide range of parameters. Sample temperatures can be varied from that of liquid nitrogen to 1700 K. Exposure to reactive gases and deposition sources can be performed simultaneously with the imaging. Finally, LEEM can be coupled with a photon source (e.g., a synchrotron) for the generation of photoemitted electron microscopy (PEEM), allowing for chemically resolved imaging. An example of the in situ, time-resolved capabilities of LEEM is the investigation of the alloy formation between tin and copper. Schmid and colleagues imaged the incorporation of tin into a copper surface, tracking the nanoscale motion of tin clusters throughout the process. [106]

Future Opportunities. Major effort must be brought to bear to improve electron optics, detectors, stage design, and computing power. The Office of Science in the Department of Energy is sponsoring an initiative to develop these advances and create the next generation of electron microscopes that will feature substantial advances in spatial, temporal, and spectral resolution concomitantly with higher brightness and sensitivity, providing unprecedented opportunities for atomic-level characterization of materials. The core of this effort will focus on overcoming the limitations currently imposed by aberrations in the electron lenses in microscopes. New technology will be developed for aberration correction that will lead to greatly improved electron beam characteristics. The benefit to the scientific community will be not only higher spatial resolution but also the ability to greatly expand the in situ environments for electron microscopy that are vital for imaging chemistry. Aberration correction will enable higher energy resolution for spectral and chemical sensitivity; faster analytical mapping; and extension of

EM techniques to chemical systems in which, previously, enough signal could not be obtained before radiation damage made the measurement irrelevant. Existing or anticipated instrumental improvements associated with aberration-compensated optics will:

- enhance cryo-tomographic studies of proteins;
- provide chemical analysis based on bonding (through valence and core-level EELS) and elemental analysis (through EELS and energy dispersive X-ray (EDX)) at very high spatial resolution (in favorable cases, atom-by-atom);
- further facilitate application of EM techniques to fully hydrated systems (e.g. environmental-EM and "wet-SEM"), thereby overcoming a major limitation of traditional, vacuum based EM.

With these new capabilities, direct imaging of chemical processes and reactions that to date have only been hypothesized will be possible. Recently, A. H. Zewail and coworkers[107] reported preliminary results on the development of 4D ultrafast electron microscopy (UEM). Providing the spatial resolution of TEM, UEM provides the ability to nondestructively image complex structures utilizing femtosecond pulses. Using UEM, it was possible to obtain images of single crystals of gold, amorphous carbon, and polycrystalline aluminum, and cells of rat intestines.

X-ray Spectroscopy and Imaging

X-rays are short-wavelength, high-energy photons. As a result, they can penetrate materials much more deeply than either visible light or electrons, producing chemical images that cannot be obtained by any other means. Chemical imaging using X-rays plays a significant role in science, health, technology development, and national security. Using X-rays, it is possible to chemically image objects as large as a shipping container or significantly smaller than the nucleus of a single cell. When X-rays interact with an atom or molecule, a variety of signals can result, depending on the type of atom and its chemical environment. Taking advantage of this phenomenon and allowing analysis of samples with wide variations in size and character have required the development of a number of different X-ray imaging techniques. These techniques can be organized into three broad categories: spectroscopy, direct imaging, and scattering. Synchrotrons produce X-rays that span a broad spectrum. For clarity, these three imaging categories have been subdivided according to the X-ray wavelength used in the experiment. Relatively long-wavelength X-rays (around 1 to 15 nm) are referred to as "soft X-rays." X-rays with wavelengths shorter than 1 nm are classified as "hard X-rays," based on the fact they are much higher in energy than soft X-rays.

Spectroscopy

By examining the energies at which X-rays are absorbed or emitted, it is possible to determine the identity and local configuration of all the atoms present in a sample. These techniques are analogous to identifying the elements involved in a combustion reaction based on the color of the flame. A single X-ray spectrum can unambiguously identify the composition and chemistry of all elements contained in a sample. The spectra also contain finer details, which can reveal chemical and magnetic information specific to the sample being studied. It is the combination of unambiguous determination of the chemistry of all elements present in the sample with subtle details of sample-specific chemical bonding that gives X-ray methods much of their utility. The spectral properties of sample constituents also form the physical basis for the contrast mechanism in direct image formation. Combining this X-ray spectral fingerprint with direct X-ray imaging is termed spectromicroscopy. This measures the local atomic-scale details of the structural and electronic environment of a chosen atomic species. Spectro-microscopy allows the fine-grained mapping of this local structure for all types of atoms throughout the entire sample volume. This provides a type of complete picture that is vital to all high-technology research and development.

Soft X-ray Spectroscopy. The unifying feature of soft X-ray spectroscopy is that some property of a sample is measured as a function of photon energy, measured either at a fixed energy or over a range of energies. Since it is possible to measure a wide range of sample-specific phenomena, many types of soft X-ray spectroscopies have been developed, including soft X-ray absorption spectroscopy (XAS), near-edge X-ray absorption fine structure (NEXAFS) spectroscopy, soft X-ray emission spectroscopy (SXES), resonant inelastic X-ray scattering (RIXS), X-ray magnetic circular dichroism (XMCD), X-ray photoemission spectroscopy (XPS), and Auger spectroscopy.

At the most basic level, the absorption, transmission, or reflectivity of a sample is measured as a function of incident photon energy. XAS and NEXAFS are methods in which the photon energy is scanned during the measurements. NEXAFS contrast soft X-ray spectromicroscopy is much better than elemental or density-based chemical imaging—it easily rivals vibrational spectroscopies with regard to functional group sensitivity. At the most sophisticated level, a "double" spectroscopy measurement can be performed. In the case of *photon-in,electron-out*, one measures the energy spectrum of photoemitted electrons (XPS) as a function of X-ray excitation energy. In the case of photon-in, photon-out, one measures the spectrum of fluorescent or inelastically scattered photons (SXES, RIXS) and does this for the range of energies of the incident photon. Often, the photon energy in XPS and SXES is chosen specifically to enhance certain absorption cross sections, but it remains fixed during the measurement. Another dimension to the technique is polarization. Certain chiral and magnetic systems respond differently to elliptical polarization produced by special synchrotron beamline insertion devices.

Soft X-ray spectroscopies employ the excitation of electrons in relatively shallow core levels (100-2000 eV) to probe the electronic structure of various kinds of matter. These techniques have been applied to imaging a number of different chemical systems, including strongly correlated materials, magnetic materials, environmental science systems, wet samples at ambient pressure, and catalysis. This type of spectroscopy relies heavily on elemental specificity. Atoms or ions of each element have their own unique set of core-level transitions that occur at characteristic energies. Through the chemical shift in both the core-level energies and the orbital energies of excited electrons, the method is sensitive to the chemical environment, such as the functional group, oxidation state, local bonding geometry, and so forth. In this sense, soft X-ray spectroscopy becomes atom specific, not just element specific. The photon energy tunability of synchrotron radiation is essential for such experiments, due to the extremely small cross sections of many atoms. Therefore, photon-in, photon-out techniques (SXES and RIXS) are viable only at bright synchrotron sources.

Hard X-ray Spectroscopy. Hard X-ray spectroscopy has been applied in a wide variety of scientific disciplines (physics, chemistry, life sciences, and geology) to investigate and image geometric and electronic structures. The method is element, oxidation, state, and symmetry specific. The technique has been particularly useful in the characterization of new materials. It has also been used in the elucidation of chemical speciation in dilute samples of environmental concern.

In the simplest experimental setup, the sample is mounted between two detectors, one that measures the incident radiation and another that measures the transmitted radiation. The ratio of incident and transmitted signals is monitored as the photon energy is swept through element-specific core-level values, or photon absorption edges. There are two main variants of the technique depending on the range of the photon energy sweep.

1. **EXAFS.** A wide sweep of the photon energy above a core-level edge displays small oscillations in the absorption from which it is possible to deduce nearest-neighbor distances and nearest-neighbor numbers. The photoelectron wave released in the absorption process bounces back to the atom of origin, not unlike the "ping" from submarine sonar. In this way, EXAFS probes the local structural and electronic environment

2. **NEXAFS.** A narrow sweep just above the core-level edge displays characteristic peaks in the spectrum that can serve as a "fingerprint" of the chemical bonding around the atom of origin.

The tunability of synchrotron radiation is essential for the sweeps across the core-level edges. The intensity of synchrotron radiation is essential for the detection of dilute species.

Direct Imaging

Direct imaging using X-rays is achieved by one of two basic experimental methods: full-field imaging or scanning. In full-field imaging, the entire sample is imaged in a single "shot." This is analogous to taking a photograph, where images are recorded on a highly pixilated X-ray-sensitive camera. With scanning imaging, a very small spot of focused X-rays is moved, or "rastered," across the sample. Images are recorded by a single-pixel detector one point at a time. These "single-point images" are then stitched together to form the full image of the sample. In either method, there has to be a physical means by which contrast is generated. This can be achieved by taking advantage of differences in absorption, refraction, composition, or the spectral properties of the chemical microenvironments being imaged. Either imaging method can, in principle, be combined with the absorption spectroscopies discussed in this section.

Soft X-ray Imaging. The nature of their interaction with matter makes soft X-rays ideal for imaging the interior structure of inorganic nanoscopic systems and biological cells. Consequently, soft X-ray microscopy has been most widely applied to chemical imaging in the fields of cell biology, environmental science, soft matter and polymers, and nanomagnetism.

Synchrotron-generated X-rays cover the entire spectral range, which allows the collection of imaging data at specific X-ray wavelengths. By a judicious choice of wavelength, it is possible to generate contrast mechanisms that differentiate between various chemical and biochemical environments. For example, in biological imaging, data are typically collected in the "water window" (300-500 eV). In this energy range, atoms such as carbon and nitrogen absorb strongly whereas water molecules are mostly transparent. This difference in absorption produces images that show striking contrast among biological structures, cellular solutions, and electron-dense markers. Similarly, in the case of nanomagnetism studies, contrast mechanisms can be generated by collecting data using magnetically sensitive, circularly polarized X-rays at the absorption wavelengths of iron, cobalt, and nickel (600-900 eV).

In recent years, a wide range of soft X-ray microscopies has been developed. These include photoelectron emission microscopy, scanning transmission X-ray microscopy (STXM), and X-ray tomography microscopy (XTM). With PEEM, a small region of the sample is illuminated and the emitted photoelectrons are passed through an EM column to produce an image. With STXM the photon energy is chosen to correspond exactly to excitations at a particular X-ray absorption resonance, leading to images with unique chemical sensitivity. Different carbon functional groups can also be detected and imaged. Vertical and horizontal polarization can be used to further enhance the image. X-ray tomography is performed by taking multiple full-field absorption images of the specimen while tilting it through 180 degrees. The three-dimensional image is then reconstructed using computer algorithms to yield quantitative information about the specimen (Figure 3.9). The resolution of STXM and TXM is on the order of 15-50 nm; that of PEEM is approximately 50 nm. Diffraction imaging microscopy (DIM) and holography are lensless imaging techniques. The future application of these techniques requires the development of ultrabright, ultracoherent sources such as the linear coherent light source (LCLS).

Hard X-ray Imaging. Similar to the way medical radiographs can reveal a broken bone, chemical imaging with hard X-rays is used to examine internal or hidden components in thick, dense samples. This technique has been applied in virtually every field from life sciences to engineering to archaeology. A few representative uses are:

- two-dimensional mapping of magnetic domains;
- three-dimensional mapping of composite materials;

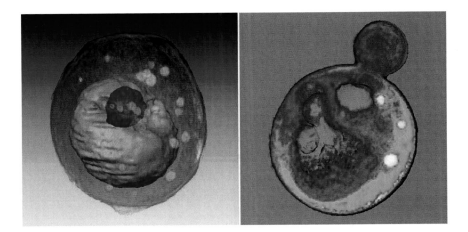

FIGURE 3.9 X-ray tomography produces images of the internal structure of cells that cannot be obtained by any other means. These images are of yeast cells, volume rendered and color coded according to the absorption of X-rays by various molecular environments in the cell. (A) The yeast nucleus (purple), vacuole (pink), and lipid droplets (white). (B) Dense lipid droplets are color-coded white, less dense vacuoles color-coded gray, and numerous other subcellular structures of intermediate densities are colored shades of green, orange, and red. Yeast cell, 5 μm diameter.
SOURCE: Courtesy of Carolyn Larabell, Lawrence Berkeley National Laboratory.

- determining the properties of individual grains in a polycrystalline material;
- mapping the distribution of elements in cells;
- identifying strains in near-perfect crystals;
- time-resolved imaging of sprays;
- investigations of human and animal physiology.

Since they have highest photon energy, hard X-rays also have the greatest depth penetration. This makes their use essential when imaging the chemistry of buried interfaces, probing the internal microstructure of objects such as bones, or imaging large biological specimens.

X-ray Scattering

Soft X-Ray Scattering. Soft X-ray scattering is a photon-in, photon-out technique. The sample is illuminated with monochromatic soft X-rays, and the scattered photons are detected over a small angular range, in either the elastic or the inelastic mode. In the former, the speckle diffraction pattern is measured, and for the latter, scattered photons are passed through a spectrometer and analyzed. The energy spectrum is essentially a replica of the occupied density of states. Additional information is obtained in the resonant condition when the incident photon is near a core-level absorption edge. Soft X-ray scattering techniques employ the excitation of electrons in relatively shallow core levels (100-2000 eV) to probe the electronic structure and other properties of various kinds of matter. The types of chemistries imaged by soft X-ray spectroscopic techniques include strongly correlated materials, magnetic materials, environmental science, wet samples at ambient pressure, and catalysis.

Every element has its own set of core levels with characteristic energies, a phenomenon that confers on these techniques a high degree of elemental specificity. Consequently, taking advantage of this can be done only by using synchrotron radiation. In addition, since many of the elements being imaged have very low absorption cross sections, many of these scattering techniques can be performed only when using the brilliant light produced at a modern, third-generation synchrotron.

Hard X-ray Scattering. When X-rays interact with matter, they are scattered. This scattering can occur in two modes, elastic or inelastic. In elastic scattering, the energy (wavelength) of the detected X-ray is the same as that of the incident X-ray. With inelastic scattering, however, some energy is lost to other processes, so the energy of the detected X-ray is lower than that of the incident X-ray. This difference in energy is due to vibrational, electronic, or magnetic excitation. The detection system in such cases measures the energy loss.

A large number of variants of hard X-ray scattering have been developed. These include small-angle X-ray scattering (SAXS), wide-angle X-ray scattering

(WAXS), grazing incidence small-angle X-ray scattering (GISAXS), X-ray Raman scattering, Compton scattering, inelastic X-ray scattering (IXS), resonant inelastic X-ray scattering, nuclear resonant scattering (NRS), and X-ray photon correlation spectroscopy (XPCS).

Hard X-ray scattering techniques have been applied to the chemical imaging of an enormous range of systems, primarily systems that are not perfectly ordered or static. Problems addressed include:

- short-range order in amorphous materials;
- liquid-vapor, liquid-liquid, and molecular film interfaces;
- colloids, solution-phase proteins, polymers;
- collective dynamics in soft materials;
- phonons and elementary excitations in solids.

Hard X-rays have wavelengths comparable to the interatomic distances. When a crystalline sample is illuminated with X-rays, the X-rays are scattered (diffracted) in very specific directions and with various specific strengths, or *intensities*. Detectors are used to measure this diffraction pattern, which is then processed to compute the arrangement of atoms within the crystal. There are two principal diffraction modes. In Bragg diffraction, the incident X-rays are monochromatic (single wavelength) and the sample is an oriented single crystal. In Laue diffraction, the incident X-ray beam is white (the entire spectrum of X-ray wavelengths) and the sample can be in the form of a powder or a random ensemble of microcrystals.

Much of the current knowledge regarding the atomic structure of materials has been derived from hard X-ray diffraction. Research problems that this technique can address are

- structural studies of crystalline materials;
- drug design by the pharmaceutical industry;
- biomineralization;
- new microporous materials including natrolites, phosphates, and titanates;
- novel complex oxides: structure-property relationships, phase transitions;
- residual stress determination in situ.

Many materials are impossible to investigate with typical research laboratory X-ray diffraction equipment, especially if the crystals are small and therefore produce diffraction intensities that are too weak to measure. Synchrotron-generated hard X-rays provide significant advantages over such laboratory sources and make these experiments possible. In this regard, one of the major areas of chemical imaging is the field of macromolecular crystallography. In particular, the technique has proven to be an exceptionally powerful tool for imaging the three-dimensional chemical environment in biological molecules and complexes. The technique has also proven amenable to applications such as drug design and ligand-binding studies, where large numbers of distinct chemical interactions must be imaged rapidly.

Macromolecular Crystallography

A major use of macromolecular crystallography is imaging potential new therapeutics in situ at the active site of an enzyme or other biomolecule. This information provides visual indications of how the potential therapeutic can be made more effective and helps cut dramatically the cost of development and the time required to produce a candidate suitable for clinical trials. In the design of a new drug, it is typical to perform a large number of cycles of imaging the drug molecule in the active site and then, based on this information, to synthesize an enhanced variant of the drug. Despite great successes, a major drawback to macromolecular crystallography is that the sample must be crystallized prior to analysis. This can often be difficult, or even impossible, since the crystallization of biomolecules remains an empirical trial-and-error process.

The structural information obtained from a crystallographic analysis has many other uses:

- elucidation of enzyme mechanisms;
- understanding structure-function relationships;
- identifying molecular recognition surfaces and topologies;
- structural genomics and proteomics;
- identification of novel "fold" motifs.

All of these implementations have enormous potential for improving health and contributing to the economy of the nation.

Summary

A plethora of X-ray chemical imaging techniques now exists. These can accommodate almost every conceivable sample type, in terms of both physical size and composition, and give unique insights into the deep internal molecular and atomic structure of most materials. On first inspection, these techniques appear complex and esoteric. The advanced capabilities of the current generation of synchrotron light sources, however, make these techniques much more accessible. Indeed, in many cases these experiments are now relatively simple and produce easily interpretable results. For example, macromolecular crystallography is now highly automated, in terms of both sample handling and postexperiment computation. At a synchrotron facility, it is possible to collect and determine the structure of a macromolecule in less than 30 minutes. This model for high-throughput imaging is being implemented on other types of experiments. The latest X-ray microscopes, especially those aimed at generating tomographic reconstructions, have been designed to function with similar speed and ease of use. Continued development of new synchrotron X-ray sources (particularly those providing soft X-rays) together with the development of plasma-based X-ray sources for use in individual research group laboratories are integral component of chemical imaging. In both soft X-ray spectromicroscopy and lensless imaging, the third-generation light sources in the United States are clearly the world leaders. Further investment could have major differential impact. The key to these areas (as in so much of imaging) is brightness. Investments in third- and fourth-generation light sources, as well as lab-based X-ray lasers and laser-excited plasmas, are likely to accelerate a chemical imaging revolution in this area. X-ray free-electron lasers have the potential to revolutionize the way we study matter at the atomic and molecular levels, allowing atomic resolution snapshots on the ultrafast time scale associated with the intrinsic motions of atoms in matter. A significant part of this task will be the development of new, enhanced X-ray optical systems such as zone plates to permit imaging at higher resolution, and the development of detectors capable of functioning on the femtosecond time scale. Recent results show that lensless imaging is capable of visualizing simple materials in isolation at approximately 10 nm resolution. This technique is applicable to specific sample geometries, (i.e., compact support).

Molecular Imaging by Mass Spectrometry

The two implementations of mass spectroscopy discussed in this section are secondary ion mass spectrometry (SIMS) and matrix-assisted laser desorption/ionization (MALDI) mass spectrometry. These techniques produce images by ionizing from a clearly identified point on a flat sample and by moving the point of ionization over the sample surface. SIMS provides information on the spatial distribution of elements and low-molecular-weight compounds as well as molecular structures of these compounds. MALDI yields spatial information about higher-molecular-weight compounds such as peptides and proteins, including their distributions in tissues at very low levels, as well as information about their molecular structures. Application of these methods to analytical problems requires appropriate instrumentation, sample preparation methodology, and data presentation usually in a three-coordinate plot where x and y are physical dimensions of the sample and z is signal amplitude. The unique analytical capabilities of mass spectrometry for mapping material and biological samples are described below.

SIMS Imaging

SIMS was the first mass spectrometry technique used to generate two-dimensional ion density maps or images from a variety of solid materials and thin sections of biological tissues.[108] SIMS involves the bombardment of a sample with a pulsed beam of ions (typically Cs^+) having energies in the low kilo-electronvolt range (10 to 15 keV). The impact of these energetic ions on the surface of the sample initiates desorption o rionization processes. The bombarding beam can be tightly focused on the surface of the sample in an area smaller than 1 μm². From a raster of the surface of the sample, ion density maps or images at specific mass-to-charge (m/z) values are generated. A wide range of various, primary ion beams are available such as Cs^+, Ga^+, Ar^+, Xe^+, and In^+. Cluster ion beams are also being actively developed including Au_n^{n+}, O_2^+, SF_5^+, C_n^+, and C_{60}^+. Cluster ion beams tend to be more efficient for desorption and ionization of higher-molecular-weight organic compounds. The primary ion energy is transferred to target atoms via atomic collisions, and a so-called collision cascade is generated. Part of the energy is transported back to the surface, allowing surface atoms and molecules to overcome the surface binding energy. The interaction of the collision cascade with surface molecules is soft enough to allow even large and nonvolatile molecules with masses up to 10,000 atomic mass units (amu) to escape with little or no fragmentation. Most of the emitted particles are neutral, but a small proportion of these are also positively or negatively charged. Subsequent mass analysis of the emitted ions provides detailed information on the elemental and molecular composition of the surface.

SIMS is a very surface-sensitive technique because the emitted particles originate from the uppermost one or two monolayers. The dimensions of the collision cascade are rather small and the particles are emitted within an area of a few nanometers' diameter. Hence, SIMS can be used for microanalysis with very high lateral resolution (50 nm to 1 μm), provided such finely focused primary ion beams can be formed. Furthermore, SIMS is destructive in nature because particles are removed from the surface. This can be used to erode the solid in a controlled manner to obtain information on the in-depth distribution of elements.[109] This dynamic SIMS mode is widely applied to analyze thin films, layer structures, and dopant profiles. To receive chemical information on the original undamaged surface, the primary ion dose density must be kept low enough ($<10^{13}$ cm^{-2}) to prevent a surface area from being hit more than once. This so-called static SIMS mode is used widely for the characterization of molecular surfaces (see Figure 3.10).

The following example of imaging with SIMS illustrates the effectiveness of using "high-mass, modest-spatial-resolution" imaging. In the analysis of photochemically produced defects on a chrome surface (Figure 3.11), imaging was performed in the focused-bunched mode of a gallium ion gun. The defects are approximately 3-5 μm in size but are easily resolved spatially with high-mass-resolution detection. The signal-to-noise ratio and acquisition time are much more efficient with this mode than with the burst mode of the gallium gun. The field of view in the images is 50.8 × 50.8 μm.

Recent developments in sample preparation for SIMS technology have incorporated the principle of using a matrix to enhance secondary ion yield.[110] In

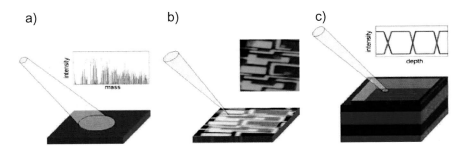

FIGURE 3.10 (a) *Surface spectroscopy*: The aim of a static SIMS investigation is analysis of the original, nonmodified surface composition. Because SIMS in principle is a destructive technique, the contribution of secondary ions to the spectrum originating from already-bombarded surface areas must be negligible. This quasi-nondestructive surface analysis can be achieved by the application of very low primary ion dose densities. (b) *Surface imaging*: By rastering a finely focused ion beam, such as an electron beam in an electron microprobe, over the surface of a sample, mass-resolved secondary ion images (chemical maps) can be obtained simultaneously. (c) *Depth profiling*: In contrast to a static SIMS experiment, high primary ion dose densities are applied, yielding successive removal (sputtering) of the respective top surface layers. By acquiring spectra during sputtering, the in-depth distribution of elements and small clusters (e.g., oxides) can be monitored. SOURCE: Copyright 2005 ION-TOF GmbH.

matrix-enhanced (ME) SIMS, the same matrices routinely in MALDI mass spectrometry (MS) are used for imaging compounds from biological tissue sections.[111] With ME SIMS, peptide ion signals can be detected and imaged in a molecular weight range up to 2500 daltons (Da) with a lateral resolution better than 3 μm. This molecular weight range is complementary to those achievable with cluster SIMS and MALDI MS.

Imaging with Laser Microprobes and MALDI Mass Spectrometry

Lasers have been used as microprobes for several decades to investigate both organic and inorganic particles.[112] More recently, laser desorption coming directly off porous silicon surfaces has been demonstrated for low-molecular-weight organic molecules, such as drugs and pharmaceutical compounds, as well as for small peptides.[113] Biological samples have also been investigated. Imaging tissue sections with laser microprobes has been demonstrated in mapping the distribution of drugs in the mouse brain, dyes from strained eye lens tissue sections, and cations in pine tree root sections.

MALDI time-of-flight (TOF) MS was introduced in the late 1980s. This technique employs the co-crystallizing of matrix (low-molecular-weight organic crys-

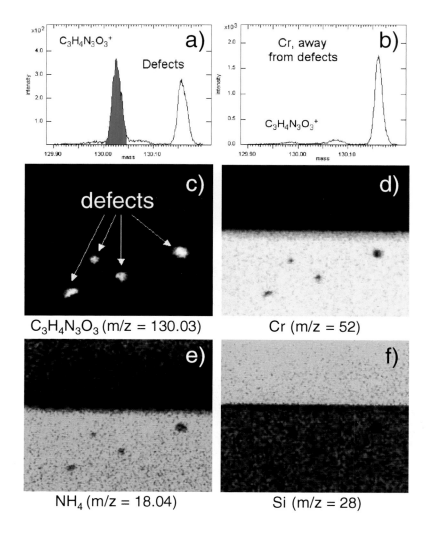

FIGURE 3.11 SIMS imaging of photochemically produced defects on a chrome surface. (a, b) High-mass-resolution spectra of defects extracted from the raw data stream of images for the $C_3H_4N_3O_3^+$ ion (m/z = 130.03). (c-f) Ion images for $C_3H_4N_3O_3^+$, Cr^+, NH_4^+, and Si^+.
SOURCE: Courtesy of Dr. J. T. Francis, Surface Science Western and Grenon Consulting, Ltd., Colchester, VT.

talline compound) and analyte on a target plate. Irradiation of these crystals by short (nanosecond time scale) pulses of UV or IR light initiates desorption and ionization, where predominantly singly protonated intact molecular ions ($[M + H]^+$) are produced. Since ionization is a pulsed process, it is easily compatible with a TOF mass analyzer. MALDI MS is an extremely sensitive tool permitting the detection of sample molecules below the femtomole level with mass accuracies better than 10^{-4} (in a mass range up to about 30 kDa). Over the past decade, many improvements to MALDI MS instrumentation have been made, and this technology is now accepted as one of the major analytical tools to detect, identify, and characterize peptides and proteins as well as many other polymers of biological interest.[114]

One of the newest developments in applications of MALDI TOF MS is its use in profiling and imaging peptides and proteins directly from surfaces such as thin-layer chromatograms and thin tissue sections in order to obtain specific information on the local molecular composition, relative abundance, and spatial distribution.[115] Results from such tissue imaging experiments yield a great wealth of information, allowing investigators to measure and compare many of the major molecular components of the section in order to gain a deeper understanding of the biomolecular processes involved. In tissue profiling experiments, one is interested in a discrete number of spots or areas in terms of comparing protein patterns. To accomplish this, matrix is homogeneously deposited on a tissue section and analyzed using an ionizing laser that is rastered over the surface of the sample according to a predetermined grid pattern of fixed dimension (Figure 3.12). The distance between two adjacent grid coordinates defines the imaging resolution. With MALDI MS, it is possible to obtain images with resolution as low as 25 μm. At each grid coordinate a full mass spectrum is recorded. Peptide and protein ion images are reconstructed by integrating the signal intensities at chosen mass values.

Beyond peptides and proteins, MALDI MS imaging of tissue section for the detection of low-molecular-weight compounds can also be achieved. Of particular interest is the posttreatment location of pharmaceutical compounds in targeted tissues or organs. Further, in parallel to location, the effects of a drug on the local proteome can be observed as a function of dose or time. Variations in the proteome are indicative of drug efficacy.[116]

Future Developments and Perspectives

Imaging by MS is currently being developed actively from both a SIMS and a MALDI perspective. Instrumentation is being upgraded constantly to perform imaging faster and at higher resolution. In SIMS MS, the development of cluster primary ion beams such as Au_n^+, Bi_n^+, and C_{60}^+ now allows the analysis of a wider range of organic molecules. ME SIMS allows us to expand the molecular weight range investigated and is valuable for the analysis of low-molecular-weight

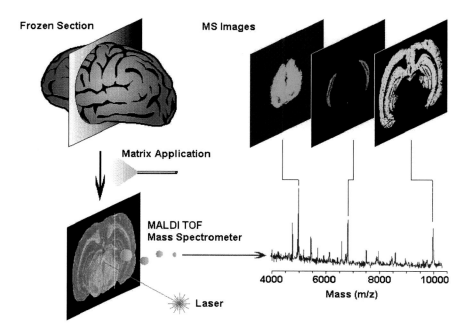

FIGURE 3.12 Principle of MALDI-based imaging mass spectrometry. Frozen sections can be mounted on a metal plate, coated with an UV-absorbing matrix, and placed in the mass spectrometer. A pulsed UV laser desorbs and ionizes analytes from the tissue, and their *m/z* values are determined using a time-of-flight analyzer. From a raster over the tissue and measurement of the peak intensities over thousands of spots, mass spectrometric images are generated at specific molecular weight values.
SOURCE: Stoeckli, M., P. Chaurand, D.E. Hallahan, and R.M. Caprioli. 2001. Imaging mass spectrometry: A new technology for the analysis of protein expression in mammalian tissues. *Nat. Med.* 7:493-496.

peptides. In parallel, ion optics allowing the better focusing of these beams are being developed. In the case of laser microprobes, the near future developments of fast lasers (with repetition rates in the kilohertz regime), improved electronics, and acquisition systems will allow significantly reduced image acquisition times (from hours to minutes) even at high resolution.

Efforts to obtain smaller laser beam size are progressing to allow imaging at higher resolution. Very accurate sample stage control systems will have to be added to the mass spectrometer ion sources. Routine imaging with laser microprobes at resolutions better than 1 micron is foreseeable within the next five years. For MALDI MS and ME SIMS, improved protocols for homogeneous matrix deposition on biological samples with the formation of micron-size crystals have to be developed. Matrix-free desorption systems allowing the study of peptides and proteins, as well as other classes of biomolecules under vacuum or at atmospheric pressure, are also needed. One such approach is already being pursued that utilizes an electrospray source as a desorption-ionization probe.[117] The rapid progression of computer technologies and informatics allows for fast data and image processing. However, with increasing acquisition and imaging resolutions, data size and volume are expected to increase significantly; thus, processing capabilities will have to keep pace with these increases. Furthermore, improved software will be needed to process and correlate images across experiments and time points. Ultimately, informatics tools will have to be developed to visualize molecular images obtained by multiple imaging technologies from the same samples.

Imaging MS is and will become increasingly critical for many aspects of materials science. One example is in the semiconductor industry, where the ability to provide spatial and chemical information on the length scales of current integrated circuit fabrication (50 nm or better) with depth profiling to provide layer-by-layer maps of the fabricated layers is critical for the continued advancement of the computer industry. Maps of any heterogeneous surface are important in other areas of materials science. For example, using various laser desorption techniques, information about the molecules found in specific inclusions in meteorites or defects in reactive surfaces can be obtained.

Molecular imaging of biological samples by MS is also foreseen to play a pivotal role in understanding numerous biological processes in fields ranging from neuroscience to cancer research. The fundamental contributions of the technology in providing molecular-weight-specific images rapidly, at relatively high resolution and sensitivity, will yield important information in the investigation of cellular processes in both health and disease. While the imaging technology can rapidly distinguish protein markers of interest, their identification is still a slow and labor-intensive process. Progress in the fields of protein identification and characterization by MS will allow a more rapid throughput. Imaging MS is of extraordinary benefit as a discovery tool because one does not need to know in advance the specific proteins that may have changed in a comparative study. For example, comparisons of protein profiles and images between tissues allow researchers to highlight protein markers indicative of the health or disease status of an individual.[118] Furthermore, the cellular origins and relative concentrations of markers across the section can be assessed, helping us understand the progression of a disease at the molecular level (e.g., cancer). Clinically, imaging mass spectrometry can provide molecular assessment of tumor staging and progression in biopsies, with the potential to identify subpopulations that are not evident based on the cellular phenotype determined histologically.[119] Also, assessment of the efficacy of drug treatment through comparative proteomics is feasible. Perhaps MS's greatest value lies in the fact that it significantly augments, but does not replace, existing molecular tools. Together, these tools promise to promote new discoveries in biology and medicine.

PROXIMAL PROBES

Since the advent of the scanning tunneling microscope in the early 1980s,[120,121] a wide variety of related microscopies using similar experimental principles and instrumentation have been developed for imaging samples based on their electronic, optical, chemical, mechanical, and magnetic properties. Today, scanning tunneling microscopy (STM)[121,122] atomic force microscopy (SFM),[122] near-field scanning optical microscopy (NSOM),[123] and scanning electrochemical microscopy (SECM)[124,125] all find broad applications in high-resolution chemical imaging experiments. Because of the similarities in the instrumentation employed in each form of microscopy, these different techniques are grouped together in this report under the general class of "proximal probe microscopies."

Imaging Modalities

By definition, proximal probe microscopes employ a small probe that is positioned very close to the sample of interest for the purposes of recording an image of the sample, performing spectroscopics experiments, or manipulating the sample. All such methods were originally developed primarily for the purposes of obtaining the highest possible spatial resolution in imaging experiments. Since then, many other unique advantages of these techniques have been realized. Proximal probe methods derive their unique capabilities specifically from the close proximity of probe and sample.

In STM, image contrast is derived from spatial variations in current flowing between the proximal probe and the sample.[126] Tunneling in an STM relies on the spatial overlap of the tip and sample electronic orbitals. Therefore, the tunneling current falls off very rapidly (on atomic length scales) as a function of distance between the tip and a particular sample feature such as an isolated atom. Tunneling current variations and information on surface chemistry are specifically derived from the associated atomic-scale variations in the density of states near the sample surface.

AFM methods derive image contrast from magnetic, electrostatic, dipolar, dispersion, and quantum mechanical (i.e., as in covalent and hydrogen bonding) interactions between the tip and sample.[127] Although magnetic, electrostatic, and dipolar interactions between the tip and sample decay over relatively long distance scales (i.e., micrometers), other forces decay over shorter angstrom ranges. Because of the short interaction lengths of the latter forces, they can be used to obtain high-resolution images of a sample surface or to make measurements of surfaces forces on isolated molecules or surface regions. AFM and variants thereof are perhaps the most widely utilized forms of scanning probe microscopy in existence today. As judged by the number of times the first article describing AFM has been cited (ranking fourth out of the top 10 published articles in *Physical*

Review Letters),[128] development of AFM represents one of the most important advances in the recent history of science.

NFOM methods (optical proximal probe methods) provide subwavelength-resolved images via the use of spatially confined evanescent fields.[129,130] These decay over length scales determined by the size of the proximal probe. High-resolution information is obtained only when subwavelength-sized probes are employed and maintained at subwavelength distances from the sample. In this near-field regime, optical fields from the proximal probe have not yet had a chance to spread through diffraction.

Scanning electrochemical microscopy relies on similar principles to obtain high resolution. In this case, however, image contrast is usually obtained via the recording of faradaic currents associated with probe and/or surface electrochemical reactions.[131] The probe must be kept small and positioned in close proximity to the sample surface to limit degradation of image resolution by diffusion of the electrochemically active species involved in generating image contrast.

In all proximal probe imaging experiments, the probe and sample are moved laterally relative to each other, and the desired signal is recorded as a function of position. In all such instruments, the resulting signal is fed into a servo (or feedback) loop for regulation of the probe-sample separation during both imaging and single-point measurements.[132]

Scanning Tunneling Microscopy and Spectroscopy

Spatial Resolution

Scanning tunneling microscopes routinely yield atomic resolution because of the extremely steep gradient of current as a function of tip-sample separation. Changes in tunneling current are readily measurable for STM probe tip motions of <0.01 Å for the most stable of such microscopes; this resolution is the result of sensitivity to the tip-sample separation. As noted above, the geometric and electronic structures are convoluted, so that rather than images of atoms, electrons distant from the nuclei (on a chemical scale) are measured with exquisite spatial resolution and significant energy resolution as well. Detailed comparison to theory is able to explain this and other features, but a general method of predicting STM images a priori remains to be developed.

STM probes (e.g., from W or Pt-It wire) are fabricated by either mechanical cutting or electrochemical etching. Further treatments are sometimes used to sharpen them, such as annealing under high fields [133,134] in techniques handed down from the original atomic resolution microscopies—field emission microscopy and field ion microscopy.[135,136] Another method employed is to lift an atom or molecule onto the probe tip so as to define the tip precisely. One important issue that can prevent clear interpretation of STM images but can also be used to tremendous advantage is the fact that the atom at the very apex of the STM probe (or rather its electronic orbitals) plays a crucial role in governing tunneling and the appearance of the sample image.[137] Figure 3.13 shows an example of this phenomenon in which Ni-O rows are imaged alternately with a clean tungsten tip and again after an oxygen atom was adsorbed to the tip apex.

Some of the most dramatic STM images have been recorded for the Si(111) 7 X 7 reconstruction, as depicted in Figure 3.14.[138] These images have been recorded at several different biases (see "Spectroscopy and Chemical Selectivity," below) and provide one of the best examples of how STM can be used to better understand the chemistry of such surfaces (specifically, the electrophilicity and nucleophilicity of the individual surface atoms). Clearly depicted in these results are the orbitals associated with surface atoms, rest atoms, and backbonds. Such studies have continued and been greatly extended into the exploration of a variety of chemical reactions that occur on silicon surfaces. These studies are described in detail in a recent review.[135]

Complete monolayers adsorbed on surfaces can also be observed by STM.[139] The separation of monolayer components in self-assembled monolayers on gold was unknown and unexpected before multicomponent films were examined using STM.[140] Having the ability to resolve components with molecular resolution,[107] this field has since advanced rapidly to exploit intermolecular interactions to produce desired patterns.[141] Such advances in other important materials could easily be driven by this ability to observe their structures and functions with atomic resolution. Other environments—liquid, gas, elevated and reduced temperature—

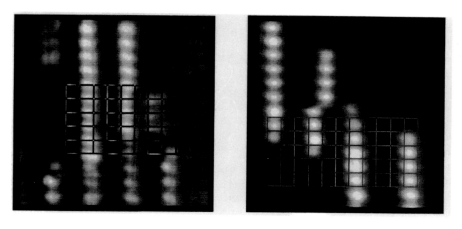

FIGURE 3.13 Ni-O rows on a Ni (1 × 1) surface imaged with a clean tungsten tip (left) and the same region after adsorption of an oxygen atom at the tip apex (right).
SOURCE: Ruan, L., F. Besenbacher, I. Stensgaard, and E. Laegsgaard. 1993. Atom resolved discrimination of chemically different elements on metal surfaces. *Phys. Rev. Lett.* 70:4079-4082.

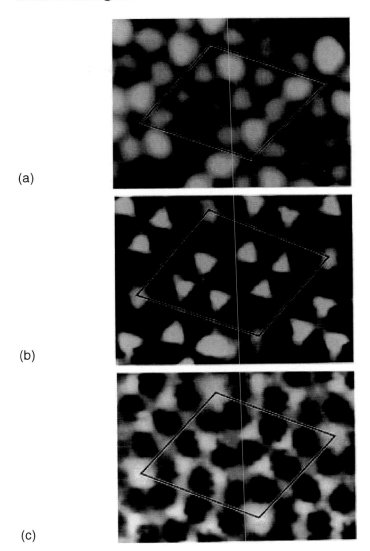

(a)

(b)

(c)

FIGURE 3.14 Using STM to understand the chemistry of surfaces. STM images of the Si(111) 7 × 7 surface reconstruction showing the spatial location of different electronic orbitals associated with different surface and near-surface atoms. The images depict the adatom states at (a) –0.35 eV, (b) –0.8 eV, and (c) –1.7 eV relative to the Fermi energy. SOURCE: Reprinted with permission from Hamers, R.J., and Y. Wang. 1996. Atomically-resolved studies of the chemistry and bonding at silicon surfaces. *Chem. Rev.* 96:1261-1290. Copyright 1996 American Chemical Society.

are all accessible, and important chemistry and materials problems await the requisite tools.

Dynamics and Time-Resolved Imaging

Not long after the invention of scanning probe microscopes, "noise" in images was attributed to the motion of atoms and molecules at rates higher than the rather slow scanning speed. The highest imaging rates are now >100 Hz,[142,143] and while some advances are possible here, much information in images is not relevant to dynamics. By focusing on the dynamic property measured, such as the location of an individual adsorbate, larger dynamic ranges can be reached.[144,145] Likewise, remaining in one position and using the STM to measure other properties also can overcome the limits imposed by mechanical scanning.

Another approach is to use pulsed or frequency-based measurements.[146] Such methods allow time resolution in the nanosecond to femtosecond regime to be achieved. This area has just begun to be explored, and as spectroscopies are further combined with scanning probes, dynamics will be made accessible by "focusing" on the relevant information.

Spectroscopy and Chemical Selectivity

Many interesting materials systems are chemically heterogeneous on a wide range of length scales down to atomic dimensions. The development of chemically selective proximal probe imaging methods has played a central role in uncovering sample heterogeneity and understanding its origins. However, numerous chemically selective, spectroscopic proximal probe methods continue to emerge from a number of labs around the world. Both the evolution of existing methods and the further development of new ones promise significant advances in our ability to obtain chemical information on heterogeneous samples on a variety of relevant length scales.

Scanning tunneling spectroscopy provides detailed information on the local electronic properties of conducting and semiconducting surfaces. The earliest tunneling spectroscopy experiments were performed by varying the bias potential applied between the proximal probe and the sample, while recording the tunneling current.[147] Such data can be recorded in a single point modality or by modulating the bias during imaging such that multiple images of the same sample region are acquired "simultaneously" at several different bias potentials. Data derived from these experiments allow the energies of both filled and unfilled electronic surface orbitals to be assessed, providing a chemically relevant view of the local surface electronic structure. It is this electronic structure that governs the chemical reactivity of such surfaces.

Feenstra and colleagues showed such chemical information in overlaying images of filled and empty orbitals on arsenic and gallium atoms of the stoichiometric GaAs(110) surface.[148] This has direct chemical implications because the electrophilic and nucleophilic centers of molecules are attracted to the filled and empty states, respectively. STM data have elucidated this effect further to show that (and which) enhancement of orbitals at specific energies is relevant to guide binding, structures, dynamics, and chemistry. Thus, STM can probe the surface the way a mobile molecule does by tuning the bias voltage to energies relevant to these interactions.[149]

As with UV-visible spectroscopy in the bulk, such techniques do not yield chemical identification, so that combining other local spectroscopies with STM is typically necessary to identify the atoms and molecules present. Specialized approaches have been developed for this, such as STM photoemission spectroscopy (PESTM) and inelastic electron tunneling spectroscopy to yield vibrational and other information. This area is extremely promising for further work in combining any number of spectroscopies with the exquisite spatial resolution of STM.

PESTM experiments are based on the fact that the tunneling process can produce electronically excited surface species that subsequently relax by radiative decay. This process is analogous to common bulk electroluminescence experiments and can be used to distinguish between chemically different surface species and/or species present in different environments. Gimzewski and coworkers and Alvarado and coworkers, both of the IBM Research Division, were the first to demonstrate PESTM experiments.[150,151]

Since its initial demonstration, PESTM has been employed in a broad range of interesting experiments, often performed as a means to better understand the local optical properties of a particular sample. An interesting recent application of PESTM is in the excitation of surface plasmon modes by inelastic tunneling processes[118] in nanostructured gold corrals.[152] PESTM imaging of the local plasmon excitation efficiency in such structures could provide a means for mapping the properties of these materials in both the presence and the absence of molecular adsorbates. Such studies could provide a means to better understand variations in enhancement factors in surface-enhanced spectroscopies employing similar substrates.[153,154]

Vibrational spectroscopy provides unique and detailed information on the chemical composition and structure of a sample. Vibrational spectra of surface-adsorbed species can also be obtained in some STM experiments using inelastic electron tunneling spectroscopy (IETS).[155] This form of spectroscopy is analogous to the sandwich tunneling junction measurements that have been made over the last 30 years.[156] Important, using IETS, this information can be obtained at the single-molecule level, with atomic-scale lateral resolution. IETS experiments are performed by measuring the tunneling conductance as a function of applied bias. When the energy of tunneling electrons matches the energy of a vibrational mode of surface-adsorbed molecules, the tunneling conductance changes as energy is deposited into the vibrations. IETS data are recorded under vacuum

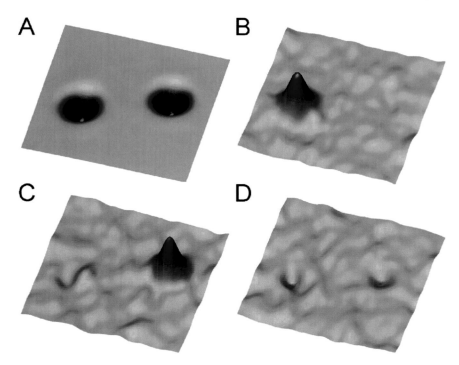

FIGURE 3.15 Spectroscopic imaging of the inelastic tunneling observed for acetylene and deuterated acetylene. (A) Regular constant-current STM image of C_2H_2 and C_2D_2 molecules (left and right). (B)-(D) Spectroscopic images recorded at 358 mV (showing C_2H_2), 266 mV (showing C_2D_2), and 311 mV, respectively.
SOURCE: Reprinted with permission from Stipe, B.C., M.A. Rezaei, and W. Ho. 1998. Single-molecule vibrational spectroscopy and microscopy. *Science* 280:1732-1735. Copyright 1998 American Association for the Advancement of Science.

conditions at cryogenic temperatures. Spectra recorded by this method can be used to identify chemical species on surfaces and to follow surface chemical reactions. Figure 3.15 show example IETS spectra obtained for acetylene and deuterated acetylene on a copper surface.[153]

Future Challenges and Emerging Methods

Like all scanning probe methods, STM experiments are limited by the rate at which images can be recorded. Again, limitations in the imaging rate arise from mechanical instrument design issues (i.e., resonances of the microscope itself). Further limitations arise from the small scan range usually employed in STM experiments. With relatively low scan rates, it is frequently difficult to image more than a tiny fraction of the actual surface area of a particular sample. Increased scanning rates might possibly be obtained in the future via the development of new feedback and scanning electronics. Improvements in both the imaging rate and the imaging area can be achieved via the implementation of multiprobe microscope designs[157,158] (i.e., with which several images can be recorded simultaneously).

Further advances will come by combination of STM with other forms of spectroscopy and scanning probe microscopy. One such emerging method that allows researchers to "see beneath the surface" of samples is ballistic electron emission microscopy (BEEM).[159]

Atomic Force Microscopy

High resolution in AFM imaging experiments is somewhat more difficult to achieve than in STM experiments. The ultimate resolution achievable in many AFM experiments is often limited by the participation of relatively long-range interactions in governing the forces between the proximal probe and sample.[160,161] Such long-range forces might arise from capillary interactions between the probe tip and the contamination layer on a sample surface imaged in the ambient environment. Relatively long-range tip-sample interactions also arise from Coulomb and dipolar interactions between probe and sample species. The shortest-range interactions result from quantum mechanical forces associated with chemical bonding. To achieve the highest possible resolution, these latter short-range forces must obviously dominate over the long-range "background" forces also present.

Along with the dominance of short-range interactions, high-resolution AFM imaging requires the use of the smallest possible probe tip. Atomic resolution imaging by definition requires the use of a probe tip with a single atom present on its apex. Pyramidal probes for AFM experiments are often obtained via a variety of well-controlled, wet-chemical and reactive ion etching procedures.[162] As such, fabrication of conventional AFM probes is extremely reproducible. Silicon nitride-based probes having nanometer-scale end diameters for use in contact and intermittent contact AFM are readily obtained from several commercial sources. Under relatively routine imaging circumstances, such probes yield resolutions in the nanometer range for relatively smooth samples.

For higher-resolution imaging and imaging of "porous" materials incorporating deep pores of narrow diameter, sharpened probes prepared by etching procedures and ion beam milling are also available. Carbon nanotube-based AFM probes have recently been developed for high-resolution imaging applications.[163] Such proximal probes have tremendous potential for use in biological imaging studies and for general materials imaging applications requiring very high aspect ratio probes.[164] In the latter case, the continued development of advanced, high-aspect-ratio probes promises to allow improved imaging of topographically complex surfaces, such as those of certain catalysts, and of porous biological membranes and other porous thin-film materials (i.e., allowing experiments to "see within the pores").

As in STM, true atomic resolution in AFM has been achieved on well-ordered, atomically smooth samples imaged under high vacuum.[165] True atomic resolution has also been demonstrated under liquids on well-ordered mineral surfaces.[166] Obtaining such high-resolution images is somewhat challenging even under high-vacuum conditions. Employment of frequency modulation AFM methods and more rigid cantilevers can improve the signal-to-noise ratio and minimize the influence of strong probe-sample interactions.

High-resolution AFM imaging is also being used to obtain a new understanding of biologically important surfaces.[167] Recent work includes the study of membrane protein arrays, such as the Aquaporin-Z protein crystals shown in Figure 3.16.[168] From repeated imaging of such proteins, an average topographical structure can be derived that can then be used along with image analysis methods to develop a detailed picture of the conformational structure of the protein. Extensive effort is also now being devoted to understanding the processes by which prion proteins aggregate and the morphological structures formed by these aggregates. Associated amyloid fibers play an important role in neurodegenerative diseases such as Alzheimer's and Creutzfeldt-Jakob disease. AFM imaging promises to have important implications for our understanding of how these diseases arise (i.e., how misfolded protein structures are "inherited" by properly folded proteins and how they might be treated).[169]

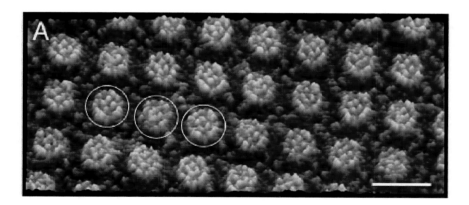

FIGURE 3.16 Example of high-resolution AFM imaging of a biological surface. Contact mode AFM image of Aquaporin-Z membrane protein crystals showing their surface structure with <1 nm resolution. Scale bar, 10 nm.
SOURCE: Horber, J.K.H., and M.J. Miles. 2003. Scanning probe evolution in biology. *Science* 302:1002-1005.

AFM Dynamics and Time Resolution

Of equal importance to achieving high spatial resolution in many proximal probe imaging experiments is the ability to acquire time-resolved images of dynamics associated with surface and thin-film chemical processes. All proximal probe methods described herein rely on the raster scanning of a single probe tip in relation to the sample of interest. The rate at which images can be recorded is the limiting factor determining the time resolution that can be obtained in dynamics experiments. Sample and/or probe raster scanning rates are limited primarily by two factors: (1) the bandwidth of the feedback circuit used for maintaining probe-sample separation and (2) the resonance frequency of the microscope.

The feedback-bandwidth limit arises primarily from the requirement that the probe follow the sample topography. In cases where the sample topography exceeds a few nanometers, imaging rates are limited to approximately one per minute or less, using conventional feedback systems. In intermittent-contact AFM and shear-force surface topographic measurements, the feedback bandwidth is limited by the response of the proximal probe cantilever (AFM) and tuning fork (shear force) to changes in the probe-sample interactions.[170] The response time in each case is often limited by the sharpness of the sensor resonance frequency (the "quality factor" or Q).[171] Higher Q values give longer response times. Improved feedback bandwidths can therefore be obtained by reducing the Q using electronic means.[172]

When the only topography is of atomic dimensions, constant height imaging methods can be employed and the feedback response is of little consequence. Under these circumstances, images can be recorded using line rates of approximately one-tenth the microscope resonance frequency, allowing tens of images to be recorded per second (i.e., at or near video rates).[173] However, faster imaging rates have also been demonstrated. In these experiments, the fast-scan motions are actually driven by one of the microscope mechanical resonances, allowing the recording of as many as 100 images per second in situations where it is not necessary that the surface topography be followed.[174]

The ultimate goal in time-resolved proximal probe methods, however, is not always *faster image acquisition*. Rather, the most useful methods provide a large dynamic range, allowing processes that occur on time scales ranging from seconds to picoseconds and even femtoseconds to be studied. Such issues are best defined in relation to the exact form of probe microscopy employed, as described below.

Chemical Force Microscopy. AFM experiments can be performed using probes that have been derivatized with specific chemical functional groups. These proximal probes allow for the detection and utilization of specific probe-sample interactions as a means of obtaining chemical contrast in AFM images, and this technique is commonly known as chemical force microscopy.[175] Chemical force microscopy represents an extrapolation to very short (i.e., nanometer) length scales of well-known surface forces measurements.[176] In these studies, probe-sample interactions are frequently detected either by measuring "pull-off" forces (force required to separate the tip and sample after contact) or by detection of frictional forces between the sample and the contacting probe.[177]

The potential for detection of chemically specific interactions between an AFM probe and a sample surface was realized using underivatized probes very early in the development of AFM, as exemplified by the detection of what appear to be discrete hydrogen bonding interactions between proximal probe and sample surface silanol species in studies by the Hansma group.[178] The capability of chemically specific AFM imaging using derivatized probes was clearly demonstrated in a later study by the Lieber group (see Figure 3.17, for example).[179]

A number of research groups have since demonstrated chemical force imaging of deliberately patterned surfaces, pointing to the broad applicability of this method to a variety of problems. These methods have been applied in force spectroscopy studies of specific single-molecule interactions (i.e., "molecule-pulling" experiments).[180] These experiments are described in the next section. The primary challenge in chemical force imaging has been in extending these procedures to imaging of samples for which the surface chemical composition is not known a priori. Although such studies can be performed on monolayer samples supported on solid substrates,[181] further difficulties arise when thicker samples (i.e., phase-separated polymer blends)[182] are to be investigated. Under these circumstances, factors such as the mechanical stiffness of the sample can lead to variations in tip-sample interaction area, making it difficult to distinguish variations in probe-sample chemical interactions. Nevertheless, researchers have recently made such measurements on a number of different samples, including oxidized polymeric surfaces.[183]

Force Spectroscopy. The atomic force microscope is also frequently used to make single-point measurements of specific interaction forces between derivatized AFM probes and sample surfaces. Tip-sample interaction forces as small as piconewtons can be measured readily. As with chemical force microscopy imaging experiments, these studies are often performed under liquids, in inert atmospheres, or in a vacuum to eliminate the strong capillary forces that can dominate tip-sample interactions.[184] Force spectroscopy (i.e., pull-off or adhesion force measurements) has been used to probe interactions between individual functional groups,[185] as in studies of interactions between carboxylic acid-terminated probe and sample surfaces.[186] Additional studies of discrete interactions between peptides and proteins have been reported, [187] and studies of discrete base-pairing (hydrogen bonding) interactions between individual nucleotide bases[188] and complementary DNA strands[189] have also been described.

Present limitations in chemical forces measurements include the need for detailed knowledge of the cantilever spring constant and the tip-sample interaction

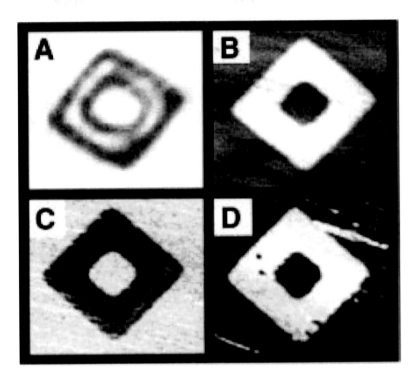

FIGURE 3.17 This image of a patterned self-assembly monolayer surrounded by a CH$_3$-terminated region was obtained using alternately a CH$_3$-terminated probe and a COOH-terminated probe in friction force and intermittent contact imaging studies. (A) Optical image showing water condensation (dark area) on the COOH-terminated region. (B) Friction force image recorded using a COOH-terminated probe. Lighter regions depict greater friction forces. (C) and (D) Tapping mode phase images of the same region using COOH- and CH$_3$-terminated probes, respectively. Darker regions depict a greater phase lag. All images are 25 µm × 25 µm.
SOURCE: Reprinted with permission from Noy, A., C.H. Sanders, D.V. Vezenov, S.S. Wond, and C.M. Lieber. 1998. Chemically sensitive imaging in tapping mode by chemical force microscopy: Relationship between phase lag and adhesion, *Langmuir* 14:1508-1511. Copyright 1998 American Chemical Society.

area in studies where quantitative values of absolute forces are to be measured. One possible solution to the problem of tip-sample contact area has been described by Beebe and coworkers.[190] In these methods, the ratio of the variance to the mean of Poisson-distributed adhesion forces yields a measure of the force associated with a single bond.

Electric Force Microscopy. Long-range forces can also be used to generate contrast in force microscopy imaging experiments. A variety of electric force measurements have been reported and variously described as scanning capacitance microscopy, Kelvin probe microscopy, and electric force microscopy.[191] Such methods have been employed to study surface charges and potential in diverse systems. Included are studies of phase separation in ionic thin films,[192] change trapping in organic semiconductor films,[193] and charge density mapping of bilayer membranes.[194] Such experiments are often limited in spatial resolution to about 50 nm, due to the long-range nature of electric forces. Implementation of carbon nanotube-based tips in electric force methods have recently been shown to yield enhanced spatial resolution.[195] Because of the long-range forces employed, electric force methods are also promising for depth-dependent sample imaging.

Surface Patterning. Patterning of surfaces using AFM and/or STM has been explored extensively in recent years.[196] Lithographic scanning probe methods involve bringing the probe near the surface to be patterned, where it can interact with and modify the local structure. The probe is then scanned laterally in a manner that will produce the desired structure. The resolution of the patterns thus produced can approach the molecular scale. Typically, changes in the surface that have been affected involve either the direct placement of molecules,[197] probe tip-mediated replacement or desorption of surface-bound molecules,[198] or probe tip-catalyzed surface reactions.[199] One limitation of many such applications is the serial nature of the fabrication process. As a result, structure fabrication is quite slow, and much effort is now being devoted to the development of parallel processing through the integration of multiple scanning probe tips.[200]

Dip-pen nanolithography (DPN) is a variety of scanning probe lithography (direct-write) developed by Mirkin and coworkers, where components of interest are transferred from an AFM tip to a substrate.[201] DPN has been used to pattern a wide variety of materials on surfaces, including small organic molecules (most commonly *n*-alkanethiols), DNA, nanoparticles, proteins, viruses, and precursors for inorganic thin films.

Liu and coworkers have used both AFM and STM to desorb molecules selectively within an alkanethiolate self-assembled monolayer (SAM).[202] The desorption mechanism differs between the two instruments: the basis of molecule removal with an AFM is detachment from the surface under an increased loading force that is significantly greater than the usual load employed for imaging. The desorption mechanism in STM lithography is electrochemical, since molecules can be desorbed by passing high-energy electrons through the film (i.e., at bias voltages of ~3-4 V).

The process of nanografting has been developed by Liu and coworkers as a method for creating both positive and negative patterns in a SAM.[203] Using AFM, molecules from a preexisting SAM matrix are removed by scanning at a force greater than the threshold displacement force. New alkanethiols are then back-filled from the contacting solution and "grafted" into the bare areas. By using longer-chain alkanethiols as the grafting solution, a positive pattern can be made; conversely, shorter-chain alkanethiols produce a negative pattern. In addition, alkanethiols possessing different functional groups (e.g., OH- or COOH-terminated) can be grafted, thereby creating a patterned SAM with varying degrees of reactivity that can be used in further applications.

Near-Field Scanning Optical Microscopy

Near-field optical microscopy involves optical imaging of a sample using a subwavelength-sized light source positioned in close proximity to the sample surface. Because NSOM employs many of the same light sources, optical elements, and detectors commonly used in conventional spectrophotometers, it is particularly simple to combine NSOM methods with those of conventional optical spectroscopy. Hence, direct spectroscopic evidence of the local chemical composition of a sample can be obtained with nanometer-scale spatial resolution using NSOM methods. Important, the data obtained from such experiments are often similar to a first approximation in form and content to those obtained in conventional far-field optical spectroscopic experiments. As a result, data interpretation is greatly facilitated via the use of well-known principles and methods for interpreting optical spectroscopic data. However, as in all proximal probe methods, NSOM images sometimes incorporate contrast arising from interactions between the probe and the sample, producing image features that are not a direct result of sample properties alone. Such probe-sample coupling poses a challenge to the interpretation of NSOM data.

To date, a number of chemically selective near-field imaging methods have been demonstrated. Near-field contrast mechanisms that rely on electronic spectroscopy (UV-visible absorption and fluorescence),[204] vibrational spectroscopy (IR absorption and Raman spectroscopies), dielectric spectroscopy (microwave dispersion), and nonlinear spectroscopy (second harmonic generation) have all been demonstrated at length scales well below the diffraction limit of light.

Fluorescence NSOM experiments have been used to observe the spatial localization of fluorescent species in composites,[205] to investigate lipid layer structures[206] and their evolution,[207] to probe biological membranes and membrane proteins,[208] to detect isolated chromophores,[209] to monitor variations in sample properties brought about by aggregation phenomena,[210] and to investigate the photophysical properties of organic semiconductors.[211] Fluorescence resonance energy transfer experiments performed using NSOM have provided particularly dramatic resolution of energy transfer processes occurring across interfaces

One of the significant promises of NSOM is the development of chemical imaging based on vibrational spectroscopic data acquired with nanometer-scale spatial resolution. The advantages of apertureless (versus aperture-based[212]) methods become particularly important in imaging experiments performed in the infrared spectral region,[213] where suitable optical fibers are difficult to obtain and implement in NSOM experiments. The apertureless approach used for the image shown in Figure 3.18 is notable because of the extremely high spatial resolution reported ($>\lambda/200$) and the absorption contrast between two polymers, imaging with laser lines that coincide with absorption bands of the polymers.

Raman spectroscopy can also be implemented in near-field imaging experiments,[214] providing an alternative method for acquiring chemically specific vibration information. Because of the extremely weak signals usually obtained in Raman experiments, Raman NSOM imaging has been greatly facilitated by the development of chemically etched probes with higher throughput.[215] Probe-enhanced Raman NSOM imaging has also been extremely important to the advancement of this method.[216] In probe-enhanced methods, the proximal probe serves to enhance the electromagnetic fields in the near-field regime in a manner similar to that of more conventional surface-enhanced Raman experiments.[217]

Although NSOM methods presently provide insufficient spatial resolution to directly resolve individual molecules in organized structures, molecular organization can still be probed using polarization-dependent NSOM methods. At present, polarization-dependent NSOM imaging can be performed only by aperture-based methods. Unfortunately, the polarization characteristics of aperture-based NSOM probes often suffer from imperfections in the metallic coating. The metal coatings on NSOM probes can be roughly on a length scale similar to the aperture size, due to grain formation. These grains alter the polarization state of the optical

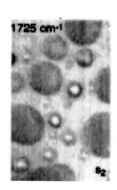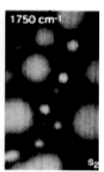

FIGURE 3.18 Images showing representative data from phase-separated domains of poly(styrene) in a poly(methyl methacrylate) matrix. *Left*: sample topography. *Right*: apertureless infrared scattering near-field optical images of poly(styrene) domains in a poly(methyl methacrylate) film recorded at the frequencies shown.
SOURCE: Reprinted with permission from Taubner, T., R. Hillenbrand, and F. Keilmann. 2004. Nanoscale polymer recognition by spectral signature in scattering infrared near-field microscopy *Appl. Phys. Lett.* 85:5064-5066. Copyright 2004, American Institute of Physics.

fields from the aperture, making interpretation of polarization-dependent images difficult. Methods for improving aperture uniformity and hence polarization quality involve focusing on ion-beam milling of the proximal end of the probes.[218]

Probing Organized Molecular Structures

Polarization-dependent and polarization-modulation NSOM imaging methods have been demonstrated for a number of organized molecular systems and for single-molecule detection.[219] Images of the fluorescence excitation (dipole) patterns of single molecules provide a dramatic view of the electric field polarization from the end of an aperture-based NSOM probe, as well as valuable information on molecular orientation (Figure 3.19). Similar methods have been used to probe molecular orientation in phospholipids films.[220] Polarization-dependent imaging studies of organized materials have also helped advance our understanding of molecular aggregates,[221] semiconducting polymers,[222] and metallic, dielectric, and inorganic semiconductor structures.[223]

Some of the most useful polarization-dependent NSOM methods, however, involve modulation of the polarization from the probe (see Figure 3.20),[224] coupled with synchronous detection of the near-field signals. Such methods allow for multiple imaging modalities so that topography, absorption dichroism, and/or birefringence information can all be readily obtained.

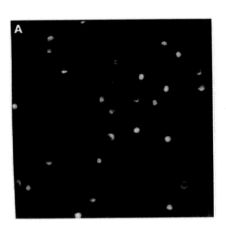 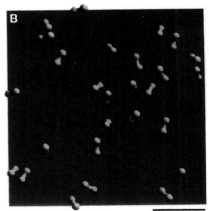

1 µm

FIGURE 3.19 (A) Dipole fluorescence excitation patterns recorded for single molecules of an indocarbocyanine dye using aperture-based NSOM methods. (B) Model of the molecular dipole orientations for the image in A.
SOURCE: Reprinted with permission from Betzig, E., and R.J. Chichester. 1993. Single molecules observed by near-field scanning optical microscopy. *Science* 262:1422-1425. Copyright 1993 American Association for the Advancement of Science.

NSOM Dynamics Imaging—Millisecond to Femtosecond Resolution

Studies of topographic and morphological dynamics by NSOM methods are usually limited to (sub)video-rate imaging, although much faster imaging has been reported under special circumstances.[225] Imaging rate limitations in NSOM not only arise from feedback bandwidth and microscope resonance issues, but also depend on optical signal levels. In transmission imaging experiments where signal levels can be significant, video-rate imaging is possible on flat (<10 nm topography) samples.[226] In other situations (i.e., fluorescence and Raman NSOM experiments), several minutes to hours are often required to acquire each image. As noted above, significant improvements in imaging rates can be achieved when high-throughput, chemically etched NSOM probes are employed.

NSOM experiments performed in the UV-visible spectral region have the unique advantage that well-developed forms of time-resolved optical spectroscopy can be implemented directly in studies of dynamics that can be retriggered at each image pixel. The dynamic range of such experiments is virtually unlimited, spanning the entire range from milliseconds[227] to femtoseconds.[228] However, in cases where fast dynamics are to be studied (i.e., nanoseconds to femtoseconds), sophisticated laser systems and electronics must be coupled with the near-field microscope. Hence, time resolution in the microsecond regime represents the practical limit for microscopes employing continuous-wave (CW) lasers and conventional optical modulators and electronics.

NSOM experiments depicting local dynamics induced by applied electric fields

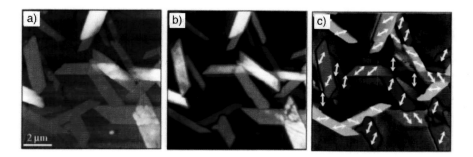

FIGURE 3.20 (a) NSOM topography, (b) polarization modulation amplitude, and (c) polarization modulation phase images of rhodamine 110 microcrystals. The arrows appended to image c depict the transition dipole orientation in each crystal.
SOURCE: Reprinted with permission from Higgins, D.A., D.A. Vanden Bout, J. Kerimo, and P.F. Barbara. 1996. Polarization-modulation near field scanning optical microscopy of mesostructured materials. *J. Phys. Chem.* 100:13794-13803. Copyright 1996 American Chemical Society.

have recently been reported and have led to better understanding of diverse processes such as charge carrier dynamics in organic semiconductor films[229] and field-induced molecular reorientation in polymer-encapsulated liquid-crystal droplets.[230]

NSOM Photolithography

Well-developed far-field photolithographic methods used routinely in the fabrication of structures tens of micrometers in size can also be employed in near-field optical microscopes for the fabrication of structures having dimensions far below the diffraction limit. In these methods, UV or visible light normally employed for photopatterning can be coupled to aperture-based probe fibers[231] or scattered from the probe tip in apertureless configurations.[232] Field confinement allows subwavelength-sized features to be produced, and fabrication of features as small as 70 nm has been reported using nonlinear excitation in an apertureless microscope.[233] Examples of what has been demonstrated to date include lithographic production of text,[234] line structures in photosentive polymers,[235] and the photochemical oxidation and removal of thiol species on gold surfaces.[236] The same limitations on lithographic speed mentioned in the discussion of AFM- and STM-based methods also apply here.

Future Challenges and Emerging Methods

Several of the existing challenges in NSOM imaging are similar to those in other scanning probe methods: namely, imaging of larger sample regions and faster image acquisition. Although parallel probe NSOM imaging is now possible,[237] it requires the use of multiple detectors and/or array detectors. Fast video-rate NSOM imaging using etched NSOM probes has also been demonstrated.[238]

An important emerging technology that promises to revolutionize near-field imaging by enhancing the spatial resolution and optical signals is the further implementation of field enhancement effects at the end of metallic apertureless probes in imaging experiments.[239] To date, field enhancement effects have been demonstrated successfully as a means to excite molecular fluorescence in the near field by both linear and nonlinear optical processes, as well as a means to enhance near-field Raman signals.[240] The design of specific proximal probe geometries that provide high field enhancements[241] for scattering and sample excitation[242] is now under way. The development of probes that are easy to fabricate and implement in NSOM experiments will increase the signals that can be obtained and hence improve imaging rates.

NSOM methods can provide extremely valuable information about the functional properties of optical and optoelectronic materials.[243] In many such applications, electrical potentials and/or electric fields are applied to samples as a means to induce changes in the local optical properties of the sample. Frequently, an electrified NSOM probe is employed in these studies.[244] The resulting changes in the sample are then detected optically in the near field. Both static and dynamic information on the variations in sample properties is thus obtained. In related methods, the probe itself or external electrodes can be used as a means to detect photocurrents[245] and/or photovoltages generated in a sample in a fashion similar to more common far-field methods. Near-field studies, however, provide the distinct advantages of higher-resolution images, coupled with invaluable topographic information that is routinely recorded along with the optical data. In the future, the chemical specificity afforded by spectroscopic NSOM methods will add a new dimension to the data, providing a direct link between local chemical composition and/or structure and materials performance.

Scanning Electrochemical Microscopy

Spatial Resolution

In SECM, a metallic probe is typically employed as a redox electrode at which an electrochemical oxidation (or reduction) process occurs. In many instances the probe-sample separation is regulated using the faradaic current as a feedback signal.[246] More recently, tuning-fork-detected shear force methods have also been used to maintain probe-sample separation,[247] opening the possibility of improved electrochemical resolution and allowing multiple simultaneous contrast mechanisms to be employed.[248] In all situations, the spatial resolution obtained in electrochemical images is limited by diffusion of redox active species between the proximal probe and the sample. Common values reported for SECM resolution are on the order of 1 μm, in cases where the probe-sample separation is large. However, resolution approaching 1 nm has been obtained in situations where a small tip is positioned in extremely close proximity to the sample.[249]

SECM imaging methods have been used to image heterogeneous surface reactivity on a number of different surfaces. Many such systems are of relevance to energy conversion and utilization. For example, SECM has been used to observe variations in the electrochemical activity of metal oxide surfaces.[250] More recently, it has been used as a means to characterize the reactivity of bimetallic catalysts for use as oxygen reduction electrodes in fuel cell systems.[251] SECM methods have also been used to image the electrochemical reactivity of biological samples, most notably living cells.[252]

Dynamics and Time-Resolved Imaging

The dynamics of charge-transfer processes are important in a wide range of biological and technological materials. SECM provides a means to study interfacial charge-transfer dynamics, while minimizing some of the difficulties associated with other methods. The rates of electron transfer can be measured by both steady-state[253] and time-resolved chronoamperometric[254] SECM methods. Mass transport through films,[255] porous membranes,[256] and biological membranes (even in living cells)[257] can also be probed. In studies of diffusion within films, methods analogous to those used in fluorescence recovery after photobleaching (FRAP) are employed to extract the desired diffusion coefficient. An initial electro-chemical potential step is employed to generate a reactant (i.e., an oxidant) at the SECM probe tip. The oxidant then locally oxidizes the film being studied. Charge migration through the film leads to the reduction of the oxidized region of the film in time. After a defined period of time, oxidant is again generated at the SECM probe, and the measured redox current provides information on the extent to which the film region has been reduced during the waiting period.

Electrochemistry and Chemical Selectivity

Scanning electrochemical microscopy is to some degree inherently chemi-cally selective, although this selectivity is not always utilized directly in SECM experiments. Chemical selectivity in electrochemical microscopy arises from the dependence of the faradaic current on the oxidation and reduction potentials of the species being detected. Examples of experiments in which the chemical selec-tivity of SECM is used to advantage include those in which transient species generated at electrode surfaces are detected. The release of certain chemicals (i.e., neurotransmitters) from cells can also be selectively detected and imaged. In one particularly interesting experiment described recently by the Mirkin group, nonmetastatic and metastatic human breast cells were imaged by SECM. Imaging was based on the measurement of redox currents arising from the products of cellular enzymatic redox reactions.[258]

Surface Patterning

SECM can be used to electrochemically desorb surface-adsorbed species (i.e., such as self-assembled monolayers)[259] or to electrochemically deposit metals[260] for the purposes of preparing surface structures of controlled geometry. The pro-duction of micrometer-sized features is accomplished by setting (or scanning) the potential of the SECM probe (or substrate) at the appropriate reducing or oxidiz-ing potential, while the probe position is scanned relative to the substrate surface. Limitations of this method include the relatively low spatial resolution that can be achieved at present and the relatively slow writing process. The resolution here again is limited by diffusion of the species being deposited or of those used to induced desorption.

Future Challenges and Emerging Methods

As with the other forms of scanning probe microscopy, SECM experiments could also be improved by the development of methods for faster image acquisition,[261] allowing better time resolution in imaging experiments. Improvements of the spatial resolution in SECM will be achieved via the continued implementation of new and/or alternative feedback mechanisms. These feedback mechanisms will rely on signals decoupled from the redox activity of the surface as a means to sense tip-sample proximity (i.e., as in the use of shear force feedback).[262] They will also allow the probe-sample separation to be maintained at extremely small values, limiting lateral diffusion of redox active species and, hence, improving spatial resolution.[263]

Recently, integration of SECM with other scanning probe techniques (STM,[264] AFM,[265] and NSOM[266]) has proved to be a valuable means for obtaining detailed, complementary chemical information. Such research should continue. Particular benefits include the ability to exploit the high-resolution imaging capabilities of STM and AFM in particular. New SECM probes will also continue to be developed. The most interesting and promising new probes presently being developed may be those based on carbon nanotubes and other nanoprobes.[267]

New developments in SECM will also include expansion of the methods available to incorporate new types of electrochemical excitation and more sophisticated data analysis. For example, the SECM probe is now being used for simultaneous amperometric and conductometric measurements, providing additional information about sample impedance on micro- and nanometer length scales.[268] Voltammetric methods, in which an entire cyclic voltammogram is recorded using the SECM probe tip at each image pixel, are also being developed for use in SECM experiments, improving their chemical selectivity.

Finally, applications of SECM to understanding the chemistry of biological systems are expanding quickly and will continue to grow. Important studies include measurements of membrane transport and characterization of ion channels and studies of charge-transfer reactions involved in photosynthesis. Imaging of single cells to better understand their function at subcellular levels will continue to expand, and the methods employed will continue to improve. One challenge in this area will be the development of SECM methods for making spatially resolved measurements inside living cells. Particularly important applications of these methods would include the development of approaches for probing the mechanisms and kinetics of biochemical reactions. This would involve the use of redox or ion-transfer mediators specific for the reaction of interest.

Emerging Methodologies

Magnetic Resonance Force Microscopy

The development of proximal probe methods by which three-dimensional images of samples can be recorded with high spatial resolution in all three dimensions would represent a major breakthrough in chemical imaging technology. Such methods would allow scientists to "see below the surface." One of the most promising emerging methods for achieving this goal is magnetic resonance force microscopy.[269] This method takes advantage of the magnetic field gradient methods employed in conventional MRI as a means to resolve sample features. It also makes use of the sensitive probe-cantilever technologies developed for more conventional atomic force imaging experiments and the ability to generate extremely strong magnetic field gradients near the end of a sharp probe.

In this experiment, a magnetic proximal probe is scanned above the surface of a sample and the resonances of electron or nuclear spins are detected as a function of position (Figure 3.21). The magnetic probe is attached to a sensitive cantilever, whose motions are detected interferometrically. As the probe is passed above a region of the sample, a change in the spin state populations beneath the probe causes a deflection of the cantilever. The ultimate sensitivity of such experiments involves detection of single electron spins, requiring field gradients of 2 gauss/nm and sensitive cantilevers and cantilever deflection methods for detecting forces as small as 2 attonewtons (aN). In typical AFM imaging experiments, the forces detected are in the piconewton to nanonewton range. To achieve single nuclear spin sensitivity, field gradients of 200 gauss/nm will be required, along with force sensitivities down to 0.3 aN.

Magnetic resonance force microscopy would allow direct chemical imaging of samples to depths of about 100 nm for a system providing 1 nm lateral resolution. In such experiments, depth resolution is obtained by scanning the radio frequency employed to flip the electron or nuclear spins in the sample.

The Photonic Force Microscope

The scanning probe methods mentioned above are predominantly surface or near-surface methods. It is difficult to obtain images of the internal features or internal surfaces of a sample using these methods. One emerging proximal probe method that may allow these limitations to be overcome is photonic force microscopy. In this method, a micrometer- or submicrometer-sized particle (a dielectric or metallic bead) is optically trapped in the focus of a laser beam. The focal position of the laser is then changed as a means of moving the particle about for imaging purposes. As the particle moves it experiences different forces associated with internal variations in the sample composition and properties. The associated forces active on the probe particle can be detected and measured by optically monitoring the position of the bead within the laser focus. Subpiconewton forces can readily be detected by this method.

The photonic force microscope may yield a means for proximal probe imaging within fluid-containing voids and structures such as vesicles and living cells. Some current limitations of the photonic force microscope include the potential for incorporation of optical artifacts when internal structures of optically complex samples (such as cells) are to be studied. Coupling of optical features into the images of such a microscope arises via the dependence of the optical trapping

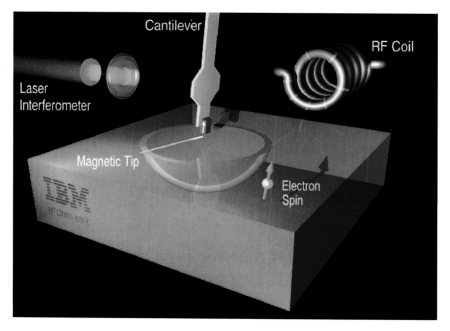

FIGURE 3.21 Magnetic resonance force microscope.
SOURCE: Figure courtesy of Dan Rugar, IBM Almaden Research Center.

potential on the optical properties of the medium through which the beam passes
in reaching the focal point.

IMAGE PROCESSING

Chemical imaging is used to selectively detect, analyze, and identify chemical and biological samples, followed by visualization of the data in the dimension
of interest. Some of the types of chemical image datasets that can be acquired are
shown in Figure 3.22. The information of interest can range from composition,
structure, and concentration to phase or conformational changes as a function of
time or temperature. The expression "chemical image" describes a multidimensional
dataset whose dimensions represent variables such as x, y, z spatial position,
experimental wavelength, time, chemical species, and so forth. Image processing
requires that the chemical images exist as digital images. A digital image is an
image $f(x,y)$ that has been digitized both in spatial coordinates and in brightness.
The value of f at any point (x,y) is proportional to the brightness (or gray level) of
the image at that point. The point at $f(x,y)$ is an image pixel, and each pixel has an
intensity value associated with it. This value may represent energy, radiation

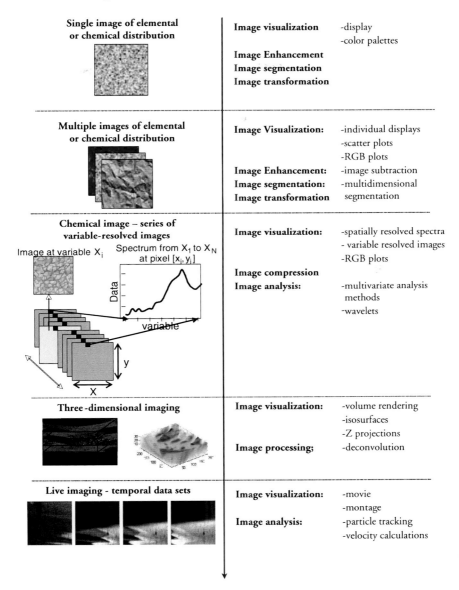

FIGURE 3.22 Integration of spatial and spectral information. NOTE: RGB = red, green, blue.
SOURCE: Julia E. Fulghum and Kateryna Artyushkova, University of New Mexico.

intensity, phase, height, et cetera, depending on the technique utilized. A digital image can be considered a matrix in which row and column indices identify a point in the image and the corresponding matrix element value identifies the gray level at that point. Image processing methods generally operate on the matrix representing the image.

Many spectroscopic imaging techniques now utilize array detectors, which allow the rapid collection of both spectral and positional data.[270] An entire spectra may be collected for each pixel on the array detector, creating a three dimensional cube consisting of both spatially resolved spectra and variable-dependent images (Figure 3.22). The complete integration of spatial and spectral information quite literally adds a new dimension to data analysis, providing the ability to examine the interdependence of spectral and spatial information, thereby improving under-standing of the underlying chemical and physical attributes. The extraction of meaningful chemical information from spectroscopic imaging datasets (multi-variate images) can require complex data analysis because more than 100 million data points can readily be acquired. Developing and implementing effective algo-rithms to obtain chemistry from spectrum imaging data is a critical and ongoing process.

Image processing is a subclass of signal processing concerned specifically with two- or three-dimensional datasets to improve image quality for human perception and/or additional mathematical analysis and interpretation.[271] Image processing is required since chemical images acquired using analytical instru-mentation can contain an overwhelming number of visual patterns generated by diverse processes. The images in an image dataset are assumed to be a composite of a chemical image and artifacts affecting image intensity that are generally unrelated to the chemical variable(s). One complicating factor in the develop-ment of image processing methods is that these artifacts are technique-specific, ranging from nonuniform illumination in optical techniques to tip artifacts in AFM.[272] One of the primary goals of image processing, regardless of the tech-nique used to generate the image, is to find a meaningful representation of the intensity distribution in a given image without introducing any artificial informa-tion by the imaging technique. That is, image processing should be carried out with caution in order to avoid excessive "beautifying" of data. The order in which image processing steps are carried out, however, and the importance of the methods mentioned here vary dramatically between imaging techniques.

A variety of methods can be used to visualize and process chemical image datasets, as listed in Table 3.1. In general, the goal is to facilitate interpretation of the dataset. In many cases the intended audience includes people that are not experts in the imaging technique, who are more interested in the chemical or spatial variation shown in the dataset. Image processing can thus have a variety of goals, including:

- enhancing contrast in images to highlight specific features;
- enhancing differences between images for correlation or comparison;
- correcting for noise, background (i.e., topography or other effects), or nonuniform illumination by removing or decreasing pixel intensities unrelated to chemical signals;
- decreasing image or dataset size to facilitate analysis (decrease analysis time) or to facilitate image correlation;
- identifying objects of interest (image segmentation, particle counting or identification);
- pre-processing (background correction, compression or binning, etc.) before utilizing multivariate analysis methods;
- quantifying image intensities;
- combining images to create (render) three-dimensional volumes (visualization).

TABLE 3.1 Various Methods to Visualize and Process Chemical Image Data Sets

Image processing category	Image processing software
Image visualization: display, color palettes, scatter plots, RGB plots	• Stretch and scale individual images with brightness and contrast • Select from a variety of different color palettes and display the associated color bar • Change lighting effects and vary surface texture • Plot image planes against one another as scatter plots • Associate up to three single-channel intensity images to produce RGB image • Assign specific colors to any number of distinct chemical or morphological species
Image enhancement: histograms, subtraction, averaging, spatial filtering and sharpening, lowpass and highpass filtering	• Compute arithmetic operations on individual images, spectra, or entire image cubes • Display image plane intensity as a histogram and use it for statistical thresholding • Perform median filtering with selective criteria • Perform image convolutions using pre-defined or user-defined kernels • Perform Savitsky-Golay smoothing with derivatives on individual spectra or entire image cubes • Spatially interpolate
Image compression: compression models, variable-length coding, JPEG and GIF standards	• Bin individual images or entire image cubes to improve signal noise ratio

continued

TABLE 3.1 Continued

Image processing category	Image processing software
Image segmentation: line and edge detection, boundary detection, thresholding, region- oriented approaches	• Create and view binary images created from image thresholding or pattern recognition techniques, and perform binary arithmetic operations • Define contours for feature extraction and automated segmentation of image planes or cubes
Image transformations: Fourier transform and spectral analysis, separable transformations, principal component analysis, wavelets; filtering techniques	• Fourier filter in the spatial (image) dimension • Deconvolve images • Image de-noising using wavelets • Spatial filtering to remove high- or low-spatial-frequency • Image compression using wavelets • Principal component analysis of multispectral image data sets with the goal of reducing dimensionality of the data, removing noise, extracting correlated information
Topological approaches: representations of boundaries and regions, morphology	• Morphologically filter images with operations such as open, close, dilate, erode, fill, clean, majority, remove, shrink, skeletonize, thicken, thin
Image restoration: noise removal	
Image recognition: statistical classifiers, neural network approaches	
Image analysis: analysis of chemically or morphologically distinct features with powerful statistical tools	• Particle size distribution calculation • Shape analysis • Granulometry analysis • Particle tracking • Texture analysis (entropy, correlation, contrast) • Roughness analysis
Multidimensional tools	• Align images within a stack • Create a montage • Create and play a movie exportable as QuickTime® or AVI • Render a HtmlResAnchor 3D reconstruction • Create HtmlResAnchor Z-series projections • View orthogonal planes • Color-combine images • Deconvolve the images • Create topographic surface maps

SOURCE: Julia E. Fulghum, University of New Mexico.

The ease with which these processes can be carried out using technique-specific software varies dramatically. Standard or add-in software on some X-ray and EM instruments can offer a range of processing options including particle counting,[273] Fourier transform-based convolution methods,[274] and tomographic image visualization,[275] while software associated with some surface analysis techniques may include only basic image manipulations such as rescaling, addition, and subtraction. Although commercial software packages exist to implement most of the processing methods described here, in general, technique-specific requirements are barriers to the average user. Image processing methods that are routine in one user community may be state of the art or unavailable in a different user community.

It is also important to be aware of the way in which image processing methods are applied to an image, so that an appropriate choice of algorithm and operating conditions can be made. For example, filters operate on groups of pixels within an image, with the number of pixels affected depending on the kernel size chosen for the filter.[276] Simple linear transformations, such as rotation or reflection, operate on single pixels, and results are independent of the value of neighboring pixels. Nonlinear transformations can dramatically alter image appearance, and Fourier transform-based methods convert images into the frequency domain for additional manipulation.

Initial Image Visualization

Frequently the first priority for the analyst is to generate an image or images that allow for visualization of heterogeneous chemical distributions in space or time. Image visualization methods vary from simply choosing a color scale for display of a single image to methods for displaying three-dimensional datasets. Simple gray-scale maps can be constructed from a single image. Different color scales can be utilized, and the contrast and brightness can be adjusted so that the information the analyst deems most important is emphasized. Multiple images from the same or different datasets can be viewed simultaneously for comparison. Scatter plots are frequently utilized for comparing two images. For more detailed comparisons among a small number of images, mapping individual images into red, green, and blue (RGB) channels creates composite color chemical images. For three-dimensional data, additional analysis tools are required, including the ability to extract spectra from a selected region of interest (ROI) for multispectral imaging datasets or rendering a three-dimensional volume or projection for depth arrays.[277]

Image Processing

A variety of factors can contribute to intensity in chemical images. Analytical microscope images can exhibit significant shading across the field of view. The shading might be caused by nonuniform illumination, nonuniform camera (detector) sensitivity, or even dirt and dust on lens surfaces if physical (rather than electrostatic or magnetic) lenses are present.[278] In surface analysis methods such as XPS and atomic emission spectroscopy (AES) the analytical signal depends not only on surface composition but also on local topography.[279] Subtraction of a background image is one of the simplest approaches to decreasing background effects, including nonuniform illumination, substrate reflectance, topography, and instrument response. This method is widely used in surface analysis methods as different as XPS and STM as well as in optically based imaging techniques.[280]

Data Extraction

A variety of tools are used for extracting image components useful in the representation or description of feature shapes. These include boundary extraction, skeletons, morphological filtering, thinning, and pruning. Image segmentation is used to separate objects of interest from the image background and is required by a variety of microscopy techniques. It is also one of the more difficult tasks in image processing.[281] Segmentation algorithms generally are based on one of two properties of the image intensity values, either discontinuities in intensity values (such as edges) or similarity according to a set of predefined criteria. Edge detection is the most common approach for detecting meaningful discontinuities in gray level.[282] An edge is defined as a set of connected pixels that lie on the boundary between two regions.

Morphological smoothing removes or attenuates both bright and dark artifacts or noise by performing an opening followed by a closing. Applying a morphological gradient highlights sharp gray-level transitions in the input image, while the top-hat transformation can be used to enhance details in the presence of shading. AFM images of biomolecules, and other structures similar in size, are enlarged because of the finite size of the probe tip.[283] A method based on morphological image processing allows analysis and correction of the enlargement.

Image transformations are operations that alter the value of pixels in an image. Transformation results do not depend on the value of neighboring pixels. These include simple linear transformations such as image rotation, translation, and reflection that may be required for correlation of images acquired using different techniques, as well as nonlinear transformations such as shearing, which is used to skew objects.

Image Analysis

The next step in image processing involves using processed images for image analysis. The ultimate aim of image analysis is generally to extract quantitative information, which may be in the form of binary presence-absence categories or of measures of object location, length or area, shape statistics, and so forth. Shape characteristics, such as area, perimeter, compactness, topological descriptors, and internal characteristics such as color and texture are among the possible types of information that may be required. Texture is used to point to intrinsic properties of a surface or region, especially those that do not have a smoothly varying intensity.[284] Texture is also used in the classification of images based on their appearance for segmentation of images into regions that have similar textural properties. Granulometry can be used to determine the number of particles, particle mean sizes, and nearest-neighbor distributions in images.[285] The results can then be correlated with macroscopic properties, such as catalytic activity. Electron microscopy (TEM, SEM) is still the most commonly used technique for visualizing colloidal particles, although scanning probe microscopy (AFM), magnetic force microscopy [MFM]) is a serious alternative that provides more information about particles.[286]

Motion analysis and particle tracking methods enable users to follow the movement over time of tagged particles, such as fluorescently labeled cell surface molecules, microtubules, nucleic acids, lipids, and other objects with subpixel resolution.[287] These methods allow scientists to measure x and y coordinates, velocity, mean displacement, mean vector length, and more.

Although, in theory, quantitative data obtained from chemical images should be useful in combination with information obtained using spectroscopic methods or bulk property measurements, image-to-property correlation is a problem poorly studied in chemical imaging. The most successful results have come from a combination of statistical image analysis, image processing, and multivariate analysis, as discussed below.

Multidimensional Image Processing

Image Compression

Image compression is an ongoing research topic in the field of image processing. A typical spectrum imaging experiment can result in the acquisition of more than 100 million data points, providing a strong impetus for automated data processing. Spectrum imaging datasets can, however, be too large to analyze efficiently, or at all, using most currently available software. For example, SIMS images are typically digitized at a resolution of 256 × 256 pixels and 16 bits of intensity information per pixel. A single image therefore requires at least 0.125 megabyte (MB) to store, and a typical image set with 64 images thus requires a minimum of 8 MB of storage. A three-dimensional SIMS analysis can involve acquisition of 4 to 8 three-dimensional datasets, resulting in 32 to 64 MB of data for only one analysis. With some types of X-ray and optical experiments, gigabyte-size datasets can result. Obviously data compression can be required to create manageable datasets.

Methods developed for traditional "lossy" image compression can be used in the compression of multispectral images. The discrete cosine transform (DCT), which is used to create standard joint photographic experts group (JPEG) images, has been widely used for lossy still image compression.[288] Although it can be efficiently implemented and performs well for high-bit-rate compression, serious blocking artifacts are a well-known disadvantage for DCT-based coding. Binning is used routinely to decrease the size of SIMS and other spectral datasets by adding intensities over a given spectral or mass range.[289] Discrete wavelet transforms (DWT) not only can overcome the blocking artifacts but also can achieve better performance overall. The wavelet transform (WT) is a new and very versatile technique that has been developed during the last decade.[290] This approach has found applications in signal and image processing[291] and, recently, also in chemistry.[292]

Developing a three-dimensional image based on an image stack can be a difficult task. In volume model visualization, information may be displayed as surfaces, interfaces, or intensity distributions through either surface or volume rendering. For automated visualization of volume data using isosurface extraction, segmentation is a necessary preprocessing step. Isosurface extraction applies a surface detector to the sample array, after which geometric primitives are fitted to the detected surfaces. These primitives are displayed using conventional surface-rendering algorithms. Isosurface extraction requires a variable contour value, and this value has a great effect on the appearance of the resulting volume.[293]

Multivariate Data Analysis Tools

Multidimensional datasets are frequently too large to analyze by visual inspection, and methods are required for reduction and analysis of these datasets. A variety of multivariate analysis (MVA) methods can be utilized to identify, extract, and correlate or classify information while reducing the overall dimensionality of the dataset.[294] MVA methods include statistical, mathematical, and graphical methods that analyze multiple variables simultaneously. MVA soft modeling methods are used for deconvolution of the original data matrix, generally including only basic, if any, physical restrictions. Physical restrictions that are sometimes utilized include nonnegativity of concentrations or molar absorptivities or the Beer-Lambert law. Principal component analysis (PCA) and classification algorithms are among the most widely applied MVA methods.[295] These methods have been successfully applied to TOF-SIMS,[296] EDX-SEM,[297] XPS,[298] IR,[299] and Raman[300] data. These methods are not available in all instrument-specific software, however, and the image preprocessing required, as well as the image processing parameters, will be strongly technique-specific. It is best to consider the MVA methods that will be utilized in initial experimental design and data acquisition because the number and type of standards utilized and the type of data acquired will strongly affect the type of multivariate analyses that can be successfully applied.

Conventional chemometric techniques such as factor analysis, least-squares fitting, PCA, and principal components regression are powerful tools for determining the composition and concentration of samples with known constituents. Neural networks and data segmentation approaches have also been successful.[301] PCA is currently one of the most commonly used methods; it transforms a number of possibly correlated variables into a smaller number of independent variables, called principal components. The first principal component accounts for as much of the variability in the data as possible, and each succeeding component accounts for a decreasing proportion of the remaining variability. The objective is to identify images that are globally correlated or anticorrelated. This information is then displayed as the loading of different images, and the pixels responsible for the correlations can be displayed in component images. The component images may be easier to interpret than pure variable images. The components can, in many cases, be connected to chemical components through a variety of methods ranging from visual inspection to additional processing such as classification or Simplisma. Principal component and cluster analysis are becoming more popular in analysis of SEM electron dispersive X-ray (EDX) data.[302] The results from PCA can be difficult to interpret. Multivariate curve resolution (MCR) is a powerful technique for extracting chemical information from multivariate images (MIs). MCR is aimed at extracting the spectra and concentrations of individual components present in mixtures using a minimum set of initial assumptions.[303] Purity-based methods show promise, and a simple, robust purity-based algorithm has been developed to initialize the MCR decomposition. Lack of selectivity, common in MI, generally results in a rotational ambiguity in factors extracted with MCR. Modifications of MCR methods are currently in development in order to reduce rotational ambiguities.[304]

In general, no software exists that is capable of visualizing multispectral imaging data, extracting those parts of the dataset that are significant, and deriving the required information. A wide range of multivariate methods have been utilized both singly and in combination, as illustrated by applications in SIMS, FTIR, SEM-EDX, Raman, NIR, and XPS.

Current Software

As discussed earlier, a significant limitation in the development of image processing is the technique-specific nature of the analysis. Several programs are available, or in development, for the application of multivariate methods. In general, these are still in the category of programs to be used by experts, because numerous decisions and assumptions must be made about data processing and interpretation. MATLAB,[305] and packages associated with it, are widely used for the application of PCA and other commonly used multivariate methods.[306] In addition to commercial add-ons, there is a user web site to which users post routines that may be of general interest. Although the posted routines can come and go, the current listing includes a graphical user interface (GUI) for visualizing three-dimensional volumetric data as well as a GUI for image segmentation and extraction.[307] A software package developed by the remote sensing community, ENVI (Environment for Visualizing Images), is used by some research groups developing methods for the analysis of chemical images.[308] ENVI combines both MVA and image visualization. A patented MCR automated software methodology has been developed for several instruments and is called expert spectral image analysis (AXSIA).[309] Although not yet publicly available, initial publications indicate AXSIA is fast enough to efficiently extract meaningful chemical components from very large spectral image datasets. The MCR methodology in AXSIA works by fitting self-generated spectral shapes to the data using least-squares procedures. Physically realizable components are obtained by applying appropriate constraints (e.g., nonnegativity of concentrations and spectra) during the solution process. The number of chemical components is estimated through an eigenanalysis of the data cross-product matrix. As an example of technique-specific data processing requirements, for analysis of TOF-SIMS images, the data are optimally scaled to account for Poisson counting statistics. This provides maximum discrimination of chemical information from noise and allows detection of small features that would be otherwise overlooked. AXSIA was tested on SEM-EDX,[310] TOF-SIMS,[311] and XPS[312] spectral imaging datasets, proving that algorithms as implemented in AXSIA are quick and efficient; they are able to process multigigabyte datasets in minutes using modern desktop computers.

Multitechnique Image Correlation

Complete characterization of a complex material requires information not only on the surface or bulk chemical components, but also on stereometric features such as size, distance, and homogeneity in three-dimensional space. It is frequently difficult to distinguish uniquely between alternative surface morphologies using a single analytical method and routine data acquisition and analysis. By combining imaging techniques that use different physical principles and, therefore, produce images representing different properties of the sample, complementary and redundant information becomes available.[313] One important goal is data fusion, which refers to combining image data from multiple techniques to form a new image or volume that contains more interpretable chemical information than can be ascertained from a single technique. Successful data fusion can decrease ambiguities in the evaluation of materials chemistry and morphology; extend lateral and vertical spatial characterization of chemical phases; or enhance spatial resolution by utilizing techniques with nanometer spatial resolution (AFM or SEM) to enhance data from techniques with spatial resolutions of microns (XPS, chemical microscopy [CM], or FTIR). This approach can facilitate correlation of different physical properties; for example correlating phase information in AFM with chemical information in XPS images.

There are MVA techniques that can used in the combination of different imaging modalities by merging registered images from different techniques. This process can also be called intermodular imaging. For example, light microscopy can be used in a variety of different imaging modes that contain complementary information. Bright-field microscopy images result from optical attenuation by the sample, whereas phase contrast microscopy images show diffractive properties. The refractive properties of the sample are displayed in differential interference contrast (DIC) images.[314] Leonardi and colleagues compared applications of polarized light, bright field, DIC, and SEM in the paper industry.[315] Fluorescence microscopy can also be correlated with images acquired using light microscopy.[316] Much effort has been expended in developing and modifying algorithms for matching images produced by different types of satellite remote sensing systems, such as optical sensors and synthetic aperture radar,[317] some of which are transferable to laboratory image correlation.

Since Raman and NIR spectroscopies are complementary in nature, their combined usage offers the opportunity to describe heterogeneous mixtures in more detail. A novel sample referencing approach has been developed that allows data to be acquired from exactly the same area of the sample using both Raman and FT-NIR microscopies. The optimum images for the components are then overlaid, which gives rise to a combined chemical image that visually describes the entire formulation. Correlating imaging XPS and imaging FTIR data from polymer blends allowed for localized quantitative studies of surface segregation and phase separation phenomena.[318] This approach is called chemical image fusion (CIF).[319]

Databasing and Data Mining

The acquisition of large datasets, followed by large numbers of such datasets, leads to issues of data cataloguing, analysis, and sharing—or databasing and data mining. Whether one is attempting to develop a method for sharing databases within a single research group or a specialized research community, numerous challenges remain in developing appropriate software for these tasks.

Databasing

Molecular databases and the associated data banks require the development of a conceptual structure for the information stored about the molecules, descriptive language representing the data, and methods for analysis enabling molecular modeling, similarity searches, classification, visualization, or other uses of the database.[320] Currently, the Protein Data Bank (PDB; *http:/www.rcsb.org/pdb/*) is one of the best known examples of a molecular database. The PDB is a worldwide archive of three-dimensional structural data of biological macromolecules.[321] The PDB is a common accentor to many structural databases.[322] The success of the PDB in enabling the statistical analysis (bioinformatics) of protein structures suggests that a broader materials image and structural database would enable similar advances (using informatics) in the understanding of generic materials.

The development of imaging databases adds additional complexities compared to molecular databases. There, is, however, significant activity in the development of software for medical images. Picture archiving and communications systems (PACS) are utilized by an increasing number of laboratories and hospitals for the storage, retrieval, and sharing of images.[323] The classical medical imaging technologies are advancing toward (1) higher resolution, (2) increased sensitivity, (3) standardized protocols, and (4) increasing application fields. The developments in items 1 to 4 will allow merging of data from different laboratories, exemplified in multimember screening studies conducted by Brown and colleagues.[324]

A database for functional magnetic resonance imaging (fMRI) provides a different example.[325] A framework that will allow scientists access to raw data from published, peer-reviewed studies has already been established (fMRI Data Center). A more demanding goal is to compile the images in a database that will allow for data mining from image sets that are both highly heterogeneous and large in size. The fMRI Data Center has adopted several guiding tenets in the organization of its core database that highlight the complexity of this task:

1. The database should be flexible enough to represent the broadest range of possible fMRI experimental paradigms.

2. The database is organized hierarchically, with the study itself at the highest level.

3. In addition to the high-level descriptive data of the study, meta-data characterization for, and pointers to, all neuroimaging data and time series are represented in the database in order to facilitate the broadest possible space over which accurate but efficient searches can be made.

4. The database should be extensible and able to incorporate new studies, scans, or time course information as it becomes available.

Clearly, the data storage issue associated with archiving functional neuroimaging data is a serious one, not to mention the computational challenges of attempting to carry out analysis on such an archive. To address these issues, establishment of "near-line" and off-line data storage is being investigated. In addition, the search to provide more suitable computing resources for carrying out large-scale analysis research is also under way.[326]

There are a variety of other databases currently in development. The Global Image Database (GID) is a web-based structural central repository (*http://www.gwer.ch/qv/gid/gid.htm*) for scientifically annotated images. The GID was designed to manage images from a wide spectrum of imaging domains ranging from microscopy to automated screening. The development of the GID is aimed at facilitating the management and exchange of image data in the scientific community and the creation of new query tools for mining image data.[327] Other databases include WebRacer, an image database that allows databasing and web serving of images; Soft Imaging System GmbH (*http://www.soft-imaging.de/rd/ english/420.htm*); and Neuroinfo (*http://www.neuroinformatica.com/faq.jsp*), a software package designed to store and serve large arrays of microscopy data. Data acquired at various magnifications can be integrated to allow navigation of the data on a number of scales.

Image and Data Mining

Image mining involves the extraction of implicit knowledge, image-data relationships, or other patterns not explicitly stored in the images or between images and other alphanumeric data. Image mining is rapidly gaining attention among researchers in the field of data mining, information retrieval, and multimedia databases because of its potential in discovering useful image patterns that may push the various research fields to new frontiers. The fundamental challenge in image mining is to determine how low-level, pixel representation contained in a raw image or image sequence can be processed efficiently and effectively to identify high-level spatial objects and relationships.

Research in image mining can be classified broadly into two main directions. The first direction involves domain-specific applications where the focus is to extract the most relevant image features into a form suitable for data mining.[328] The second direction involves general applications where the focus is to generate image patterns that maybe helpful in understanding the interaction between high-level human perceptions of images and low-level image features.[329] The latter may lead to improvements in the accuracy of images retrieved from image databases.

Image mining is not simply an application of existing data mining techniques to the image domain because there are important differences between relational databases and image databases:

1. *Absolute versus relative values.* In relational databases, the data values are often readily interpretable. For example, age is 35 is well understood. However, in image databases, the data values (e.g., pixel intensities) have a significance that will depend on the context. For example, a gray-scale value of 46 could appear darker than a gray-scale value of 87 if the surrounding context pixels values are all very bright.

2. *Spatial information.* Another important difference between relational databases and image databases is that implicit spatial information is critical for interpretation of image contents, but there is no such requirement in relational databases. One approach to this problem is to extract position-independent features before searching for patterns between image datasets.

3. *Unique versus multiple interpretations.* A third important difference is associated with the fact that multiple interpretations may apply to the same visual patterns observed in images. Traditional data mining algorithms, which associate a data pattern with a specific class or interpretation, are less useful for analysis of images. A new class of discovery algorithms is needed in response to the requirements for mining useful patterns from images.

The image database containing raw image data cannot be used directly for mining purposes. Raw image data have to be processed to generate information that is usable for high-level mining modules. An image mining system is often complicated because it employs various approaches and techniques ranging from image retrieval and indexing schemes to data mining and pattern recognition. A good image mining system is expected to provide users with effective access into the image repository and generation of knowledge and patterns underneath the images. Such a system typically encompasses the following functions: image storage, image processing, feature extraction, image indexing and retrieval, patterns, and knowledge discovery. Two different frameworks can be used to distinguish image mining systems: function-driven and information-driven image mining frameworks. The function-driven framework focuses on the functionalities of different component modules to organize image mining systems, while the latter is a hierarchical structure with an emphasis on the information needs at various levels in the hierarchy. Image mining techniques include object recognition, image indexing and retrieval, image classification and clustering, association rules mining, and neural network.[330]

Generating Images Through Theory and Simulation (Cyberinfrastructure)

Image databasing and data mining generally refer to the compilation and analysis of experimentally acquired images. Effective methods for exploring information obtained in chemical images acquired in the lab versus those developed through theory and simulation are also insufficiently developed. When one speaks of chemical imaging of samples, a subtle distinction can be drawn between (1) the creation of an image of a *particular* sample and (2) the creation of an image of a *generic* sample. In the former, there is no recourse but to make a literal measurement of the *particular* sample of the various types outlined in the other parts of this chapter. In this case, the role of cyberinfrastructure is primarily image manipulation as discussed above, although the attempts to address or understand a *particular* sample may be aided by whatever techniques have been used to understand *generic* samples. In the latter case, however, cyberinfrastructure and its underlying theoretical frameworks can provide computer-generated images that provide insight into the structure and function of *generic* samples. Moreover the simulations can also allow test cases for new paradigms for the cybertools used to manipulate data, process metrics, and render images using the experimental measurements.

Theoretical Models or Representations. The 10^{10} range of magnitude from the atomic scale to the bench top requires a significant amount of averaging (or compressing) to remove nonessential data. On the one hand, all the data cannot be stored easily, but on the other hand, it is a significant undertaking to search through all of the data. At present, this means pursuing multiscale techniques. One such hierarchical approach uses small-scale models to generate parameters for the next-scale model, and so forth. The lowest-scale system (at the molecular level) can be described by a combination of quantum mechanics and classical mechanics. It is the success of such models that has led to the utility and widespread use of molecular dynamics simulations. Yet what does one do at higher scales—perhaps use effective classical particles or integrative representations? Both are being done, although the methods are still in their infancy.

Yet another promising line of research lies in creating mesoscopic representations whose fundamental scale is somewhere within the 10^{10} range of distance scales and for which one defines closed (or fully consistent) equations of motion. At the macroscopic limit, hydrodynamic models are a very successful and standard example. More recent approaches include the Cahn-Hillard coarse-grained models and phase-field models. In some cases, one aims to ascertain the degree to which the systems exhibit self-similarity at various length scales; hence the lack of a specific parameterization—which would be necessary using reduced-dimensional models—is not of much importance.

Computer Simulation

To image either possible structures or trajectories of a system, one specifies the fundamental (smallest-length scale) representation, the interactions between objects in the system, and the appropriate equations of motion for this representation. When this fundamental representation is at the atomistic level, there is a wide array of choices for the interactions, ranging from the most computationally expensive, using high levels of ab initio calculations, to the least expensive, using parameterized force fields. At present, one can routinely run reasonable simulations consisting of about 100,000 atoms for up to a nanosecond using a few days of computational time. This is sufficient to obtain structural and dynamic information for many systems. However, this approach does not readily provide dynamic information about complex processes such as chemical reactions and full protein folding events. The latter have been treated, at present, with a variety of simplifying approximations or algorithms. For example, the solvent may be represented using mean-field forces, thereby removing the detailed descriptions of the solvent molecules. This is a common and elementary example of hierarchical models in which different regions of the system are treated with varying degrees of accuracy.

At another extreme, one can use effective mesoscopic models to characterize a given system and, in the best cases, obtain accurate structural information at much larger length scales than these found in atomic simulations. Such multi-scaling techniques can be used to perform simulations for length scales as large a centimeters and for time scales as long as hours. However, they sacrifice detailed information about length scales shorter than micrometers.

Regardless of how the simulation is performed, the results must be analyzed. At present, only a limited number of cybertools are available for measuring the properties of simulations beyond merely recording the trajectories. Examples of such cybertools found routinely in many packages involve the calculation of correlation functions, triangulation of structure, Fourier transforms, clustering metrics, and informatics-based metrics.

Molecular Dynamics Simulations

At the atomic and molecular scale, typical simulation techniques use molecular dynamics (MD) to integrate the classical equations of motion.[331] MD simulations are particularly desirable for current computational platforms because they are often highly parallelizable and limited only by the many-body terms in the force fields. Modern computers are sufficiently fast that many simulations can be run in real time to observe the molecular motions. Nonetheless asynchronous computing to obtain larger or longer MD simulations can readily be performed, often utilizing the same computing platforms. The necessity for such large-scale simulations lies in the fact that many processes cannot be isolated to a few interacting molecules, and they often require an explicit representation of the molecular environment. One extreme example using embarrassingly parallel computing is that of the Folding@Home project.[332] Several packages are now available that simplify the process of implementing these algorithms to arbitrary systems, such as DL_POLY,[333] and a module in NWCHEM.[334] A larger number have been written specifically for biological systems such as CHARMM,[335] TINKER,[336] GROMACS,[337] and NAMD.[338] In the lowest order of accuracy, the force fields are generally pairwise, but they accommodate the largest systems for the longest trajectories. Nevertheless, as higher levels of accuracy are required, most modern potentials implemented in the cited codes also include higher-order corrections, including multipoles and polarization effects. In addition, so-called transferable force fields are increasingly being developed, allowing investigators a larger palette of molecules and larger portions of the phase diagram.[339] For cases in which the force fields are unavailable or the underlying electron quantization is important, then Born-Oppenheimer Molecular Dynamics (BOMD), [340] Carr-Parrinello Molecular Dynamics (CPMD),[341] or Atom-Centered Density Matrix Propagation (ADMP)[342] have been integrated into commercial and free-ware computer codes.

In summary, at present there exist several cybertools for performing MD simulations of various systems by experts or near experts. These tools are fairly mature in capability, but the user interfaces have only recently started to make them available to scientists other than the experts in the computational chemistry community. The challenge will be to improve these cybertools to make their use completely transparent.

Multiscaling Simulations

While MD techniques are somewhat mature and have led to many successes in modeling the structure and dynamics of molecular systems, the use of multi-scale methods connecting this microscopic level and the macroscopic world is growing now.[343] Indeed the extensive set of contributions (more than 3500 pages) in the recent compendium[344] describing multiscaling techniques serves to illustrate the importance of bridging this gap, as well as the breadth of techniques that are being aimed at it. Currently there exist many theories and a large effort in developing cybertools; however, the current state of the art does not contain mature computational packages at the level of the existing MD simulation packages. The development of such packages (and the associated advances in theoretical methods) in the context of addressing visualization of structures and dynamics is clearly an emerging area that would help advance chemical imaging.

Future Directions

Several promising avenues of research would greatly enhance the technological capacity of image processing, simulation, and modeling. To expedite the development of these various technologies, alterations in the field's current landscape are necessary. First, more accurate algorithms are needed for all imaging applications to ensure that the results generated from the algorithm are legitimate for varying implementations of the algorithm. One of the most critical needs associated with this task is a concerted effort to determine which algorithms are best suited for a particular image processing method. In addition, a more effective transfer of computer science advances in general image processing to scientific fields is urgently needed. To facilitate this transfer, rigorous collaborations between computer scientists and researchers using chemical imaging will be required. Furthermore, there is a serious gap in knowledge between the experts and routine users in the chemical imaging field. Often, researchers are unaware of or unfamiliar with which image processing methods, particularly multivariate methods, are most useful. If these rapid, target-specific routines could be incorporated into standard technique software it would accelerate widespread appropriate use in the field. There is also a need for investigators to incorporate the use of a priori information in order to optimize data analysis and image processing conditions. This requirement stems from the fact that many users apply image processing methods as though they know nothing about the sample(s) or system(s) when, in reality, users are rarely operating blindly. A related consideration is that most computer software is not optimized for speed or management of large datasets; researchers are therefore often discouraged from trying different approaches to the analysis of a single dataset. Increases in computational capacity are thus essential to generate novel image processing methods. Finally, to enhance our understanding of a variety of chemical processes, visualization methods must be developed and improved. Such methods are currently lagging behind image processing methods, and developed visualization methods have been implemented slowly in some chemical imaging communities. Developing visualization methods that correlate data across length and time scales and have the capacity to display and analyze such data will be crucial to the advancement of the field. Meanwhile, an ability to predict and describe structures at various levels using computer models is a powerful tool to help guide the visualization and interpretation of particular samples. For example, this ability could help bridge gaps in missing information or insufficient resolution (as long as it is used carefully.) Alternatively, it could help guide experimentalists to identify what regions of a structure would be particularly interesting. Thus, although such modeling would not image a particular structure, it is an extremely valuable tool that could help the overall effort.

NOTES AND REFERENCES

1. The committee would like to thank the following individuals for contributions made to this chapter: Steven R. Higgins, Wright State University; Takashi Ito, Kansas State University; Michael V. Mirkin, Queens College, City University of New York (CUNY); Rachel K. Smith, Pennsylvania State University; David O. Wipf, Mississippi State University; Kateryna Artyushkova, University of New Mexico; and Svitlana Pylypenko, University of New Mexico.

2. *For the Ernst and Wuthrich Nobel lectures, see* http://nobelprize.org/chemistry/laureates/1991/ernst-lecture.html *and* http://nobelprize.org/chemistry/laureates/2002/wuthrich-lecture.html.

3. Rugar, D., R. Budakian, H.J. Mamin, and B.W. Chui. 2004. Single spin detection by magnetic resonance force microscopy. *Nature* 430:329-332.

4. Greenberg, Y.S. 1998. Application of superconducting quantum interference devices to nuclear magnetic resonance. *Rev. Mod. Phys.* 70:175-222.

5. Hu, K. N., H.H. Yu, T.M. Swager, and R.G. Griffin. 2004. Dynamic nuclear polarization with biradicals. *J. Am Chem Soc.* 126:10844-10855.

6. Han, S.I., S. Garcia, T.J. Lowery, E.J. Ruiz, J.A. Seeley, L. Chavez, D.S. King, D.E. Wemmer, and A. Pines. 2005. NMR-based biosensing with optimized delivery of polarized ^{129}Xe to solutions. *Anal. Chem.* 77:4008-4012.

7. Godard, C., S. B. Duckett, S. Polas, R. Tooze, and A.C. Whitwood. 2005. Detection of intermediates in cobalt-catalyzed hydroformylation using para-hydrogen-induced polarization. *J. Am. Chem. Soc.* 127:4994-4995.

8. Bax, A. 2003. Weak alignment offers new NMR opportunities to study protein structure and dynamics. *Protein Sci.* 12:1-16.

9. Tugarinov, V., J.E. Ollerenshaw, and L.E. Kay. 2005. Probing side-chain dynamics in high molecular weight proteins by deuterium. *J. Am. Chem. Soc.* 127:8214-8225.

10. Tycko, R. 2001. Biomolecular solid state NMR: Advances in structural methodology and applications to peptide and protein fibrils. *Annu. Rev. Phys. Chem.* 52:575-606.

11. For the Lauterbur and Mansfield Nobel lectures, see *http://nobelprize.org/medicine/laureates/2003/lauterbur-lecture.html* and *http://nobelprize.org/medicine/laureates/2003/mansfield-lecture.html.*

12. Moonen, C.T.W., and P.A. Bandettini, eds. 1999. *Functional MRI.* Berlin: Springer-Verlag.

13. Gillard J., A. Waldman, and P. Barer, eds. 2005. *Clinical MR Neuroimaging: Diffusion, Perfusion, Spectroscopy.* Cambridge, U.K.: Cambridge University Press.

14. *http://www.magnet.fsu.edu.*

15. *http://www.meteoreservice.com/neurospin/.*

16. Mcdermott, R., S.K. Lee, B. Kahem, A.H. Trabesinger, A. Pines, and J. Clarke. 2004. Microtesla MRI with a superconducting quantum interference device. *Proc. Natl. Acad. Sci. USA* 101:7857-7861.

17. de Zwart, J.A., P.J. Ledden, P. van Geldern, J. Bodurka, R. Chu, and J.H. Duyn. 2004. Signal-to-noise ratio and parallel imaging performance of a 16-channel receive-only brain coil array at 3.0 Tesla. *Magn. Resonan. Med.* 51:22-26.

18. Sodickson, D.K., and C.A. McKenzie. 2001. A generalized approach to parallel magnetic resonance imaging. *Med. Phys.* 28:1629-1643.

19. Hu, K.N., H.H. Yu, T.M. Swager, and R.G. Griffin. 2004. Dynamic nuclear polarization with biradicals. *J. Am Chem Soc.* 126:10844-10855.

20. Golman, K., L.E. Olsson, O. Axelsson, S. Mansson, M. Karlson, and J.S. Petersson. 2003. Molecular imaging using hyperpolarized ^{13}C. *Br. J. Radiol.* 76:S118-127.

21. Han, S.I., S. Garcia, T.J. Lowery, E.J. Ruiz, J.A. Seeley, L. Chavez, D.S. King, D.E. Wemmer, and A. Pines. 2005. NMR-based biosensing with optimized delivery of polarized ^{129}Xe to solutions. *Anal. Chem.* 77:4008-4012.

22. Albert, M.S., G.D. Cates, B. Driehuys, W. Happer, B. Saam, C.S. Springer, and A. Wishnia, A. 1994. Biological magnetic resonance imaging using laser-polarized ^{129}Xe. *Nature* 370:199-201.

23. Rugar, D., R. Budakian, H.J. Mamin, and B.W. Chui. 2004. Single spin detection by magnetic resonance force microscopy. *Nature* 430:329-332.

24. Jaffer, F.A., and R. Weissleder. 2005. Molecular imaging in the clinical arena. *JAMA* 293:855-862.

25. Michaels, C.A., S.J. Stranick, L.J. Richter, and R.R. Cavanagh. 2000. Scanning near-field infrared microscopy and spectroscopy with a broadband laser source. *J. Appl. Phys.* 88:4832.

26. (a) Haka, A.S., K.E. Shafer-Peltier, M. Fitzmaurice, J. Crowe, R.R. Dasari, and M.S. Feld. 2005. Diagnosing breast cancer using raman spectroscopy. *Proc. Natl. Acad. Sci.* 102:12371-12376.

(b) Widjaja, E., T.-C. Chen, M.D. Morris, M.A. Ignelzi, Jr., and B.R. McCreadie. 2003. Band-target entropy minimization (BTEM) applied to hyperspectral Raman image data. *Appl. Spectrosc.* 57:1353-1362.

(c) Caspers, P.J., G.W. Lucassen, and G.J. Puppels. 2003. Combined in vivo confocal Raman spectroscopy and confocal microscopy of human skin. *Biophys. J.* 85:572-580.

27. (a) Jeanmaire, D.L., and R.P.V. Duyne. 1977. Surface Raman spectroscoelectrochemistry. Part 1. Heterocyclic, aromatic, and aliphatic amines adsorbed to anodized silver electrodes. *J. Electroanal. Chem.* 84:1-20.

(b) Albrecht, M.G., and J.A. Creighton. 1977. Anomalously intense Raman spectra of pyridine at a silver electrode. *J. Am. Chem. Soc.* 99:5215-5217.

28. Moskovits, M. 1985. Surface-enhanced spectroscopy. *Rev. Mod. Phys.* 57:783-826.

29. Capadona, L.P., J. Zheng, J.I. González, T.-H. Lee, S.A. Patel, and R.M. Dickson. 2005. Nanoparticle-free single molecule anti-Stokes Raman spectroscopy. *Phys.l Rev. Lett.* 94):058301.

30. Adams, D.M., L. Brus, C.E.D. Chidsey, S. Creager, C. Creutz, C.R. Kagan, P.V. Kamat, M. Lieberman, S. Lindsay, R.A. Marcus, R.M. Metzger, M.E. Michel-Beyerle, J.R. Miller, M.D. Newton, D.R. Rolison, O. Sankey, K.S. Schanze, J. Yardley, and X.Y. Zhu. 2003. Charge transfer on the nanoscale: Current status. *J. Physi. Chem. B* 107:6668-6697.

31. Kneipp, K., W. Yang, H. Kneipp, I. Itzkan, R.R. Dasari, and M.S. Feld. 1996. Population pumping of excited vibrational states by spontaneous surface-enhanced Raman scattering. *Phys. Rev. Lett.* 76:2444-2447.

32. Nie, S., and S.R. Emory. 1997. Probing single molecules and single nanoparticles by surface-enhanced Raman scattering. *Science* 275:1102-6.

33. Shen, Y.R. 1984. *The Principles of Nonlinear Optics.* New York: John Wiley & Sons.

34. Duncan, M.D., J. Reintjes, and T.J. Manuccia. 1982. Scanning coherent anti-Stokes Raman microscope. *Opt. Lett.* 7:350.

35. Zumbusch, A., G.R. Holtom, and X.S. Xie. 1999. Three-dimensional vibrational imaging by coherent anti-Stokes Raman scattering. *Phys. Rev. Lett.* 82:4142.

36. Cheng, J.-X., and X.S. Xie. 2004. Coherent anti-Stokes Raman scattering microscopy: Instrumentation, theory, and applications. *J. Phys. Chem. B* 108:827.

37. Potma, E.O., and X.S. Xie. 2004. CARS microscopy for biology and medicine. *Opt. Photon. News* 15:40.

38. Hartschuh, A., E.J. Sanchez, X.S. Xie, and L. Novotny. 2003. High-resolution near-field Raman microscopy of single-walled carbon nanotubes. *Phys. Rev. Lett.* 90:095503.

39. Weissleder, R. 2001. A clearer vision for in vivo imaging. *Nat. Biotech.* 19:316-317.

40. (a) Docherty, F.T., M. Clark, G. McNay, D. Graham, and W.E. Smith. 2004. Multiple labelled nanoparticles for bio detection. *Faraday Dis.* 126:281-288.

(b) Cao, Y.W.C., R.C. Jin, and C.A. Mirkin. 2002. Nanoparticles with Raman spectroscopic fingerprints for DNA and RNA detection. *Science* 297:1536-1540.

41. Kneipp, J., H. Kneipp, W.L. Rice, and K. Kneipp. 2005. Optical probes for biological applications based on surface-enhanced Raman scattering from indocyanine green on gold nanoparticles. *Anal.l Chem.* 77:2381-2385.

42. (a) Bhargava, R., and I.W. Levin. 2003. Time-resolved Fourier transform infrared spectrosocpic imaging. *Appl. Spectrosc.* 57:357-366.

(b) Chan, K.L.A., S.G. Kazarian, A. Mavraki, and D.R. Williams. 2005. Fourier transform infrared imaging of human hair with a high spatial resolution without the use of a synchrotron. *Appl. Spectrosc.* 59:149-155.

(c) OuYang, H., P.J. Sherman, E.P. Paschalis, A. Boskey, and R. Mendelsohn. 2004. Fourier transform infrared microscopic imaging: Effects of estrogen and estrogen deficiency on fracture healing in rat femurs. *Appl.Spectrosc.* 58:1-9.

43. For a comprehensive overview of recent developments in infrared imaging, note an upcoming book, *Spectrochemical Analysis Using Infrared Multichannel Detectors,* edited by Rohit Bhargava and Ira Levin (Blackwell Publishing, 2006).

44. Kneipp, J., L.M. Miller, M. Joncic, M. Kittel, P. Lasch, M. Beekes, and D. Naumann. 2003. In situ identification of protein structural changes in prion-infected tissue. *Bioch. Biophys. Acta* 1639:152-158.

45. Williams, G.P. 2004. High power THz synchrotron sources. *Phil. Trans. R. Soc. Lond.* A 362:403.

46. Globus, T., D. Woolard, M. Bykovskaia, B. Gelmont, L. Werbos, and A. Samuels. 2003. THz-frequency spectroscopic sensing of DNA and related biological materials. *Int. J. High Speed Electron. Syst.* 13:903-936.

47. Shaner, N.C., R.E. Campbell, P.A. Steinbach, B.N.G. Giepmans, A.E. Palmer, and R.Y. Tsien. 2004. Improved monomeric red, orange and yellow fluorescent proteins derived from *Discosoma* sp. red fluorescent protein. *Nat. Biotech.* 22:1567-1572.

48. Haupts, U., S. Maiti, P. Schwille, and W.W. Webb. 1998. Dynamics of fluorescence fluctuations in green fluorescent protein observed by fluorescence correlation spectroscopy. *Proc. Nat. Acad. Sci. USA* 95:13573-13578.

49. Pologruto, T.A., R. Yasuda, and K. Svoboda. 2004. Monitoring neural activity and Ca^{2+} with genetically encoded Ca^{2+} indicators. *J. Neurosci.* 24:9572-9579.

50. Thompson, R.E., D.R. Larson, and W.W. Webb. 2002. Precise nanometer localization analysis for individual fluorescent probes. *Biophys. J.* 82:2775-2783.

51. Michalet, X., F.F. Pinaud, L.A. Bentolila, J.M. Tsay, S. Doose, J.J. Li, G. Sundaresan, A.M. Wu, S.S. Gambhir, and S. Weiss. 2005. Quantum dots for live cells, in vivo imaging, and diagnostics. *Science* 307:538-544.

52. Yao, J., D.R. Larson, W. R. Zipfel, and W.W. Webb. 2005. Dark fraction and blinking of water-soluble quantum dots in solution. *Proc. SPIE: Nanobiophoton. Biomed. Appl. II,* Vol. 5705, Cartwright, A. N., and M. Osinski, eds. Bellingham, WA: SPIE Press.

53. Ow, H., D.R. Larson, M. Srivastava, B.A. Baird, W.W. Webb, and U. Wiesner. 2005. Bright and stable core-shell fluorescent silica nanoparticles. *Nano Lett.* 5:113-117.

54. (a) Magde, D., E. Elson, and W.W. Webb. 1972. Thermodynamic fluctuations in a reacting system—Measurement by fluorescence correlation spectroscopy. *Phys. Rev. Lett.* 29:705-708.

(b) Webb, W.W. 2001. Fluorescence correlation spectroscopy: Inception, biophysical experimentations, and prospectus. *Appl. Opt.* 40:3969-3983.

55. Foquet, M., J. Korlach, W.R. Zipfel, W.W. Webb, and H.G. Craighead. 2002. DNA fragment sizing by single molecule detection in submicrometer-sized closed fluidic channels. *Anal. Chem.* 74:1415-1422.

56. (a) Xie, X.S., and J.K. Trautman. 1998. Optical studies of single molecules at room temperature. *Annu. Rev. Phys.l Chem.* 49:441-480.

(b) Moerner, W.E., and M. Orrit. 1999. Illuminating single molecules in condensed matter. *Science* 283:1670-1676.

(c) Ishijima, A., and T. Yanagida. 2001. Single molecule nano-bioscience. *Trends Biochem. Sci.* 26:438-444.

57. Lu, H.P., L. Xun, and X.S. Xie. 1998. Single-molecule enzymatic dynamics. *Science* 282:1877-1882.

58. Stryer, L., and R.P. Haugland. 1967. Energy transfer: A spectroscopic ruler. *Proc. Natl. Acad. Sci. USA* 58:719-726.

59. Weiss, S. 2000. Measuring conformational dynamics of biomolecules by single molecule fluorescence spectroscopy. *Nat. Struct. Biol.* 7:724-729.

60. Zhuang, X., H. Kim, M.J.B. Pereira, H.P. Babcock, N.G. Walter, and S. Chu. 2002. Correlating structural dynamics and function in single ribozyme molecules. *Science* 296:1473-1476.

61. Levene, M.J., J. Korlach, S.W. Turner, M. Foquet, H.G. Craighead, and W.W. Webb. 2003. Zero-mode waveguides for single-molecule analysis at high concentrations. *Science* 299:682-686.

62. Ashkin, A., J.M. Dziedzic, J.E. Bjorkholm, and S. Chu. 1986. Observation of a single-beam gradient force optical trap for dielectric particles. *Opt. Lett.* 11:288.

63. Strick, T., J.F. Allemand, V. Croquette, and D. Bensimon. 2000. Twisting and stretching single DNA molecules. *Prog. Biophys. Mol. Bio.* 74:115.

64. Visscher, K., M.J. Schnitzer, and S.M. Block. 1999. Single kinesin molecules studied with a molecular force clamp. *Nature* 400:184-189.

65. (a) Bustamente, C., Y.R. Chemla, N.R. Forde, and D. Izhaky. 2004. Mechanical processes in biochemistry. *Annu. Rev. Biochem.* 73:705-748.

(b) Allemand, J.F., D. Bensimon, and V. Croquette. 2003. Stretching DNA and RNA to probe their interactions with proteins. *Curr. Opin. Struct. Biol.* 13:266-274.

66. Yildiz, A., J.N. Forkey, S.A. McKinney, T. Ha, Y.E. Goldman, and P.R. Selvin. 2003. Myosin V walks hand-over-hand: Single fluorophore imaging with 1.5 nm localization. *Science* 300(5628):2061-2065.

67. Klar, T.A., S. Jakobs, M. Dyba, A. Egner, and S.W. Hell. 2000. Fluorescence microscopy with diffraction resolution barrier broken by stimulated emission. *Proc. Natl. Acad. Sci. USA* 97:8206-8210.

68. White, J.G., W.B. Amos, and M. Fordham. 1987. An evaluation of confocal versus conventional imaging of biological structures by fluorescence light-microscopy. *J. Cell Biol.* 105:41-48.

69. Pawley, J.B. 1995. *Handbook of Biological Confocal Microscopy (The Language of Science)*. New York: Plenum.

70. Denk, W., J.H. Strickler, and W.W. Webb. 1990. Two-photon laser scanning fluorescence micrscopy. *Science* 248:73-76.

71. Holtmaat, A., J.T. Trachtenberg, L. Wilbrecht, G.M. Shepherd, X.Q. Zhang, G.W. Knott, and K. Svoboda. 2005. Transient and persistent dendritic spines in the neocortex in vivo. *Neuron* 45:279-291.

72. Pologruto, T.A., R. Yasuda, and K. Svoboda. 2004. Monitoring neural activity and Ca^{2+} with genetically encoded Ca^{2+} indicators. *J. Neurosci.* 24:9572-9579.

73. Maiti, S., J.B. Shear, R.M. Williams, W.R. Zipfel, and W.W. Webb. 1997. Measuring serotonin distribution in live cells with three-photon excitation. *Science* 275:530-532.

74. (a) Jung, J.C., and M.J. Schnitzer. 2003. Multiphoton endoscopy. *Opt. Lett.* 28:902-904.

(b) Levene, M.J., D.A. Dombeck, R.P. Molloy, K.A. Kasischke, R.M. Williams, W.R. Zipfel, and W.W. Webb. 2004. In vivo multiphoton microscopy of deep brain tissue. *J. Neurophysiol.* 91:1908-1912.

(c) Levene, M.J., D.A. Dombeck, R.M. Williams, J. Skoch, G.A. Hickey, K.A. Kasischke, R.P. Molloy, M. Ingelsson, E.A. Stern, J. Klucken, B.J. Backskai, W.R. Zipfel, B.T. Hyman, and W.W. Webb. 2004. In vivo multiphoton microscopy of deep tissue with gradient index lenses. *Photon. West 2004.* San Jose: International Society for Optical Engineering.

75. Helmchen, F., M.S. Fee, D.W. Tank, and W. Denk. 2001. A miniature head-mounted two-photon microscope: High-resolution brain imaging in freely moving animals. *Neuron* 31:903-912.

76. Kasischke, K.A., H.D. Vishwasrao, P.J. Fisher, W.R. Zipfel, and W.W. Webb. 2004. Neural activity triggers neuronal oxidative metabolism followed by astrocytic glycolysis. *Science* 305:99-103.

77. Fields, R.D. 2004. The other half of the brain. *Sci. Am.* 290:54-61.

78. Vishwasrao, H.D., A.A. Heikal, K.A. Kasischke, and W.W. Webb. 2005. Conformational dependence of intracellular NADH on metabolic state revealed by associated fluorescence anisotropy. *J. Biol. Chem.* 280:25119-25126.

79. Dombeck, D.A., M. Blanchard-Desce, and W.W. Webb. 2004. Optical recording of action potentials with second-harmonic generation microscopy. *J. Neurosci.* 24:999-1003.

80. (a) Freund, I., and M. Deutsch. 1986. 2nd-harmonic microscopy of biological tissue. *Opt. Lett.* 11:94-96.

(b) Mertz, J., and L. Moreaux. 2001. Second-harmonic generation by focused excitation of inhomogeneously distributed scatterers. *Opt. Comm.* 196:325-330.

(c) Williams, R.M., W.R. Zipfel, and W.W. Webb. 2005. Interpreting second harmonic generation images of collagen I fibrils. *Biophys. J.* 88:1377-1386.

81. Barad, Y., H. Eisenber, M. Horowitz, and Y. Silberberg. 1997. Nonlinear scanning laser microscopy by third harmonic generation. *Appl. Phys. Lett.* 70:922-924.

82. (a) Moreaux, L., O. Sandre, and J. Mertz. 2000. Membrane imaging by second-harmonic generation microscopy. *J. Opt. Soc. Am. B* 17:1685-1694.

(b) Cheng, J.-X., and X.S. Xie. 2002. Greens function formulation for third-harmonic generation microscopy. *J. Opt. Soc. Am. B* 19:1604-1610.

(c) Cheng, J.-X., A. Volkmer, and X.S. Xie. 2002. Theoretical and experimental characterization of coherent anti-Stokes Raman scattering microscopy. *J. Opt. Soc. Am. B* 19:1363-1375.

83. Stubbs, C.D., S.W. Botchway, S.J. Slater, and A.W. Parker. 2005. The use of time-resolved fluorescence imaging in the study of protein kinase C localization in cells. *BMC Cell Biol.* 6:22.

84. Bowen, B.P., J. Enderlein, and N.W. Woodbury. 2003. Single-molecule fluorescence spectroscopy of TOTO on poly-AT and poly-GC DNA. *Photochem.Photobiol.* 78(6):576-581.

85. Uversky, V.N., S. Winter, and G. Lober. 1996. Use of fluorescence decay times of 8-ANS-protein complexes to study the conformational transitions in proteins which unfold through the molten globule state. *Biophys. Chem.* 60(3):79-88.

86. (a) Wan, S.K., Z.X. Guo, S. Kumar, J. Aber, and B.A. Garetz. 2004. Noninvasive detection of inhomogeneities in turbid media with time-resolved log-slope analysis. *J. Quant. Spectros. Radiat. Transfer* 84:493-500.

(b) Bordenave, E., E. Abraham, G. Jortusauskas, J. Oberle, and C. Rulliere. 2002. Single-shot correlation system for longitudinal imaging in biological tissues. *Opt. Comm.* 208:275-283.

87. Inoue, S. 1986. *Video Microscopy.* New York: Plenum Publishing Corporation.

88. Fujimoto, J.G., C. Pitris, S.A. Boppart, and M.E. Brezinski. 2000. Optical coherence tomography: An emerging technology for biomedical imaging and optical biopsy. *Neoplasia* 2:9-25.

89. See for example:

(a) Dragsten, P.R., W.W. Webb, J.A. Paton, and R.R. Capranic. 1974. Auditory membrane vibrations—measurements at sub-angstrom levels by optical heterodyne spectroscopy. *Science* 185:55-57.

(b) Denk, W., and W.W. Webb. 1990. Optical measurement of picometer displacements of transparent microscopic objects. *Appl. Opt.* 29:2382-2391.

90. Gustafsson, M.G.L. 1999. Extended resolution fluorescence microscopy. *Curr. Opin. Struct. Biol.* 9:627-634.

91. See for example:

(a) Sinclair, M.B., J.A. Timlin, D.M. Haaland, D.M., and M. Werner-Washburne. 2004. Design, construction, characterization, and application of a hyperspectral microarray scanner. *Appl. Opti.* 43:2079-2089.

(b) Schultz, R.A., T. Nielsen, J.R. Zavaleta, R. Ruch, R. Wyatt, and H.R. Garner, 2001. Hyperspectral imaging: A novel approach for microscopic analysis. *Cytometry* 43:239-247.

(c) Zimmerman, T., J. Rietdorf, J. and R. Pepperkok. 2003. Spectral imaging and its applications in live cell microscopy. *Fed. Europ. Biochem.Soc. Lett.* 546: 87-92.

92. Furuta, T., S.S.H. Wang, J.L. Dantzker, T.M. Dore, W.J. Bybee, E.M. Callaway, W. Denk, and R.Y. Tsien. 1999. Brominated 7-hydroxycoumarin-4-ylmethyls: Photolabile protecting groups with biologically useful cross sections for two photon photolysis. *Proc. Natl. Aacd. Sci. USA* 96:1193-1200.

93. Chklovskii, D. B., B.W. Mel, and K. Svoboda. 2004. Cortical rewiring and information storage. *Nature* 431:782-788.

94. (a) Zipfel, W.R., R.M. Williams, R. H. Christie, A.Y. Nikitin, B.T. Hyman, and W.W. Webb. 2003. Live tissue intrinsic emission microscopy using multiphoton excited intrinsic fluorescence and second harmonic generation. *Proc. Natl. Acad. Sci. USA* 100:7075-7080.

(b) Zipfel, W.R., R.M. Williams, and W.W. Webb. 2003. Nonlinear magic: Multiphoton microscopy in the biosciences. *Nat. Biotech.* 21:1369-1377.

95. Ouzounov, D.G., K.D. Moll, M.A. Foster, W.R. Zipfel, W.W. Webb, and A.L. Gaeta. 2002. Delivery of nanojoule femtosecond pulses through large-core microstructured fibers. *Opt. Lett.* 27:1513-1515.

96. (a) Leapman, R.D. 2004. Novel techniques in electron microscopy. *Curr. Opin. Neurobiol.* 14:591-598.

(b) Frank, J. 2002. Single-particle imaging of macromolecules by cryo-electron microscopy. *Ann. Rev. Biophys. Biomol. Struct.*31:303-319.

97. Rossmann M.G., M.C. Morais, P.G. Leiman, and W. Zhang. 2005. Combining X-ray crystallography and electron microscopy. *Structure* 13:355-362.

98. Rodgers, D.W. 2001. Cryocrystallography techniques and devices. Pp. 202-208 in *International Tables for Crystallography*, Volume F: *Crystallography of Biological Macromolecules*, M.G. Rossmann and E. Arnold, eds. Dordrecht: Kluwer Academic Publishers.

99. (a) Baker, T.S., N.H. Olson, and S.D. Fuller. 1999. Adding the third dimension to virus life cycles: Three-dimensional reconstruction of icosahedral viruses from cryo-electron micrographs. *Microbiol. Mol. Biol. Rev.* 63:862-922.

(b) Dubochet, J., M. Adrian, J.J. Chang, J.C. Homo, J. Lepault, A. W. McDowall, and P. Schultz. 1988. Cryo-electron microscopy of vitrified specimens. *Q. Rev. Biophys.* 21:129-228.

100. (a) Tao, Y., and W. Zhang. 2000. Recent developments in cryoelectron microscopy reconstruction of single particles. *Curr. Opin. Struct. Biol.* 10:127-136.

(b) van Heel, M., B. Gowen, R. Matadeen, E.V. Orlova, R. Finn, T. Pape, D. Cohen, H. Stark, R. Schmidt, and M. Schatz. 2000. Single particle electron cryo-microscopy: Towards atomic resolution. *Q. Rev. Biophys.* 33:307-369.

101. (a) Henderson, R. 2004. Realizing the potential of electron cryomicroscopy. *Q. Rev. Biophys.* 37:3-13.

(b) van Heel, M., B. Gowen, R. Matadeen, E.V. Orlova, R. Finn, T. Pape, D. Cohen, H. Stark, R. Schmidt, and M. Schatz. 2000. Single particle electron cryo-microscopy: Towards atomic resolution. *Q. Rev. Biophys.* 33:307-369.

102. *http://www.ornl.gov/info/press_releases/get_press_release.cfm?ReleaseNumber=mr20040917-00.*

103. Klie, R.F., and N.D. Browning. 2000. Atomic scale characterization of a temperature dependence to oxygen vacancy segregation at $SrTiO_3$ grain boundaries. *Appl. Phys. Lett.* 77:3737-3739.

104. Klie, R.F., and Y. Zhu. 2005. Atomic resolution STEM analysis of defects and interfaces in ceramic materials. *Micron* 36 (3):219-231.

105. Williamson, M.J., R.M. Tromp, P.M. Vereecken, R. Hull, and F.M. Ross. 2003. Dynamic microscopy of nanoscale cluster growth at the solid-liquid interface. *Nat. Mat.* 2:532-536.

106. Schmid, A.K., N.C. Bartelt, and R.Q. Hwang. 2000. Alloying at surfaces by the migration of reactive two-dimensional islands. *Science* 290:1561-1564.

107. Lobastov, V. A., R. Srinivasan, A. H. Zewail. 2005. Four-dimensional ultrafast electron microscopy. *Proc. Natl. Acad. Sci.* 102):7069-7073.

108. (a) Benninghoven, A. 1994. Surface-analysis by secondary-ion mass-spectrometry (SIMS). *Surf. Sci.* 300:246-260.

(b) Chabala, J.M., K.K. Soni, J. Li, K.L. Gavrilov, and R. Levisetti. 1995. High-resolution chemical imaging with scanning ion probe SIMS. *Int. J. Mass Spectom. Ion Process.* 143:191-212.

109. Wucher, A., S. Sun, C. Szakal, and N. Winograd. 2004. Molecular depth profiling in ice matrices using C-60 projectiles. *Appl. Surf. Sci.* 231-232:68-71.

110. McDonnell, L.A., T.H. Mize, S.L. Luxembourg, S. Koster, G.B. Eijkel, E. Verpoorte, N.F. de Rooij, and R.M. A. Heeren. 2003. Using matrix peaks to map topography: Increased mass resolution and enhanced sensitivity in chemical imaging. *Anal. Chem.* 75: 4373-4381.

111. McDonnell, L.A., S.R. Piersma, A.F.M. Altelaar, T.H. Mize, S.L. Luxembourg, P. Verhaert, J. van Minnen, and R.M.A. Heeren. 2005. Subcellular imaging mass spectrometry of brain tissue. *J. Mass Spectrom.* 40:160-168.

112. (a) Hercules, D.M., F.P. Novak, S.K. Viswanadham, and Z.A. Wilk. 1987. Applications of laser microprobe mass-spectrometry in organic-analysis. *Anal. Chim. Acta* 195:61-71.

(b) Seydel, U., M. Haas, E.T. Rietschel, and B. Lindner. 1992. Laser microprobe mass-spectrometry of individual bacterial organisms and of isolated bacterial compounds—A tool in microbiology. *J. Microbiol. Methods* 15:167-183.

(c) Vaeck, L.V., H. Struyf, W.V. Roy, and F. Adams. 1994. Organic and inorganic analysis with laser microprobe mass spectrometry. Part I: Instrumentation and methodology. *Mass Spectrom. Rev.* 13:189-208.

113. Bhattacharya, S.H., T.J. Raiford, and K.K. Murray. 2002. Infrared laser desorption/ionization on silicon. *Anal. Chem.* 74:2228-2231.

114. (a) Pandey, A., and M. Mann. 2000. Proteomics to study genes and genomes. *Nature* 405:837-846.

(b) McDonald, W.H., and J.R. Yates. 2000. Proteomic tools for cell biology. *Traffic* 1:747-754.

(c) Aebersold, R., and D.R. Goodlett. 2001. Mass spectrometry in proteomics. *Chem. Rev.* 101:269-295.

(d) Godovac-Zimmermann, J., and L.R. Brown. 2001. Perspectives for mass spectrometry and functional proteomics. *Mass Spectrom. Rev.* 20:1-57.

115. (a) Stoeckli, M., P. Chaurand, D.E. Hallahan, and R.M. Caprioli. 2001. Imaging mass spectrometry: A new technology for the analysis of protein expression in mammalian tissues. *Nat. Med.* 7:493-496.

(b) Gusev, A.I., O.J. Vasseur, A. Proctor, A.G. Sharkey, and D.M. Hercules. 1995. Imaging of thin-layer chromatograms using matrix-assisted laser desorption/ionization mass spectrometry. *Anal. Chem.* 67:4565-4570.

116. Reyzer, M.L., R.L. Caldwell, T.C. Dugger, J.T. Forbes, C.A. Ritter, M. Guix, C.L. Arteaga, and R.M. Caprioli. 2004. Early changes in protein expression detected by mass spectrometry predict tumor response to molecular therapeutics. *Cancer Res.* 64:9093-9100.

117. Takats, Z., J.M. Wiseman, B. Gologan, and R.G. Cooks. 2004. Mass spectrometry sampling under ambient conditions with desorption electrospray ionization. *Science* 306:471-473.

118. Chaurand, P., S.A. Schwartz, and R.M. Caprioli. 2004. Assessing protein patterns in disease using imaging mass spectrometry. *J. Proteome Res.* 3:245-252.

119. Chaurand, P., M.E. Sanders, R.A. Jensen, and R.M. Caprioli. 2004. Proteomics in diagnostic pathology: Profiling and imaging proteins directly in tissue sections. *Am. J. Pathol.* 165:1057-1068.

120. Binnig, G., and H. Rohrer. 1982. Scanning tunneling microscopy. *Helv. Phys. Acta* 55:726-735.

121. Binnig, G., H. Rohrer, C. Gerber, and E. Weibel. 1982. Surface studies by scanning tunneling microscopy. *Phys. Rev. Lett.* 49:57-60.

122. Binnig, G., C.F. Quate, and C. Gerber. Atomic force microscope. *Phys. Rev. Lett.* 56:930-933.

123. (a) Synge, E.H. 1928. A suggested method for extending microscopic resolution into the ultra-microscopic region. *Phil. Mag.* 6:356-362.

(b) Ash, E.A., and G. Nicholls. 1972. Super-resolution aperture scanning microscope. *Nature* 237:510-512.

(c) Lewis, A., M. Isaacson, A. Harootunian, and A. Muray. 1984. Development of a 500 angstrom spatial resolution light microscope. *Ultramicroscopy* 13:227-232.

(d) Pohl, D.W., W. Denk, and M. Lanz. 1984. Optical stethoscopy: Image recording with resolution $\lambda/20$. *Appl. Phys. Lett.* 44:651-653.

124. Bard, A.J., F.-R. Fan, J. Kwak, and O. Lev. 1989. Scanning electrochemical microscopy: Introduction and principles. *Anal. Chem.* 61:132-138.

125. Engstrom, R.C., M. Weber, D.J. Wunder, R. Burgess, and S. Winquist. 1986. Measurements within the diffusion layer using a microelectrode probe. *Anal. Chem.* 58:844-848.

126. (a) Chen, C.J. 1993. *Introduction to Scanning Tunneling Microscopy*. New York: Oxford University Press.

(b) Wiesendanger, R. 1994. *Scanning Probe Microscopy and Spectroscopy: Methods and Applications*. New York: Cambridge University Press.

(c) Bonnell, D.A. 2001. *Scanning Probe Microscopy and Spectroscopy*. New York:Wiley-VCH.

127. (a) Ibid.

(b) Sarid, D. 1994. *Scanning Force Microscopy*. New York: Oxford University Press.

128. Riordan, J. 2003. *Physical Review Letters top ten*. *APS News* 12(5):3,6.

129. Paesler, M.A., and P.J. Moyer. 1996. *Near-Field Optics: Theory, Instrumentation, and Applications*. New York: John Wiley & Sons.

130. Dunn, R.C. 1999. Luminescence in scanning tunneling microscopy on III-V nanostructures. *Chem. Rev.* 99:2891-2927.

131. Mirkin, M.V., and A.J. Bard, eds. 2001. *Scanning Electrochemical Microscopy*. New York: Marcel Dekker.

132. (a) Chen, C.J. 1993. *Introduction to Scanning Tunneling Microscopy*. New York: Oxford University Press.

(b) Wiesendanger, R. 1994. *Scanning Probe Microscopy and Spectroscopy: Methods and Applications.* New York: Cambridge University Press.

(c) Bonnell, D.A. 2001. *Scanning Probe Microscopy and Spectroscopy.* New York:Wiley-VCH.

(d) Sarid, D. 1994. *Scanning Force Microscopy.* New York: Oxford University Press.

(e) Paesler, M.A., and P.J. Moyer. 1996. *Near-Field Optics: Theory, Instrumentation, and Applications.* New York: John Wiley & Sons.

(f) Mirkin, M.V., and A.J. Bard, eds. 2001. *Scanning Electrochemical Microscopy.* New York: Marcel Dekker.

133. Binh, V.T., and N. Garcia. 1991. Atomic metallic ion emission, field surface melting, and scanning tunneling microscopy tips. *J. Phys. I* 1:605-612.

134. Binh, V.T., and N. Garcia. 1992. On the electron and metallic ion emission from nanotips fabricated by field-surface-melting technique: Experiments on W and Au tips. *Ultramicroscopy* 42-44:80-90.

135. Hren, J.J., and J. Liu. 1995. Field emission microscopy. In *The Handbook of Surface Imaging and Visualization,* Hubbard, A.T., ed. New York: CRC Press.

136. Ehrlich, G., and N. Ernst. 1995. Field ion microscopy and spectroscopy. In *The Handbook of Surface Imaging and Visualization,* Hubbard, A.T., ed. New York: CRC Press.

137. Ruan, L., F. Besenbacher, I. Stensgaard, and E. Laegsgaard. 1993. Atom resolved discrimination of chemically different elements on metal surfaces. *Phys. Rev. Lett.* 70:4079-4082. See also note 115 (a).

138. Hamers, R.J., and Y. Wang. 1996. Atomically-resolved studies of the chemistry and bonding at silicon surfaces. *Chem. Rev.* 96:1261-1290.

139. (a) Poirier, G.E. 1997. Characterization of organosulfur molecular monolayers on Au(111) using scanning tunneling microscopy. *Chem. Rev.* 97:1117-1127.

(b) Giancarlo, L.C., and G.W. Flynn. 1998. Scanning tunneling and atomic force microscopy probes of self-assembled, physisorbed monolayers: Peeking at the peaks. *Annu. Rev. Phys. Chem.* 49:297-336.

(c) Giancarlo, L.C., and G.W. Flynn. 2000. Raising flags: Applications of chemical marker groups to study self-assembly, chirality, and orientation of interfacial films by scanning tunneling microscopy. *Acc. Chem. Res.* 33:491-501.

(d) De Feyter, S., A. Gesquiere, M.M. Abdel-Mottaleb, P.C.M. Grim, F.C. De Schryver, C. Meiners, M. Sieffert, S. Valiyaveettil, and K. Mullen. 2000. Scanning tunneling microscopy: A unique tool in the study of chirality, dynamics, and reactivity in physisorbed organic monolayers. *Acc. Chem. Res.* 33:520-531.

140. Stranick, S.J., A.N. Parikh, Y.-T. Tao, D.L. Allara, and P.S. Weiss. 1994. Phase separation of mixed-composition self-assembled monolayers into nanometer scale molecular domains. *J. Phys. Chem.* 98:7636-7646.

141. (a) Liu, G.-Y., S. Xu, and Y. Qian. 2000. Nanofabrication of self-assembled monolayers using scanning probe lithography. *Acc. Chem. Res.* 33:457-466.

(b) Kramer, S., R.R. Fuierer, and C.B. Gorman. 2003. Scanning probe lithography using self-assembled monolayers. *Chem. Rev.* 103:4367-4418.

(c) Smith, R.K., P.A. Lewis, and P.S. Weiss. 2004. Patterned self-assembled monolayers. *Prog. Surf. Sci.* 75:1-68.

142. Frenken, J.W.M., T.H. Oosterkamp, B.L.M. Hendriksen, and M.J. Rost. 2005. Pushing the limits of SPM. *Mat. Today* 8:20-25.

143. Rost, M.J., L. Crama, P. Schakel, E. van Tol, G.B.E.M. van Velzen-Williams, C.F. Overgauw, H. ter Horst, H. Dekker, B. Okhuijsen, M. Seynen, A. Vijftigschild, P. Han, A.J. Katan, K. Schoots, R. Schumm, W. van Loo, T.H. Oosterkamp, and J.W.M. Frenken. 2005. Scanning probe microscopes go video rate and beyond. *Rev. Sci. Instrum.* 76:053710.

144. Pohl, D.W., and R. Moller. 1988. Tracking tunneling microscopy. *Rev. Sci. Instrum.* 59:840-842.

145. Swartzentruber, B.S. 1996. Direct measurement of surface diffusion using atom-tracking scanning tunneling microscopy. *Phys. Rev. Lett.* 76:459-462.

146. (a) Hamers, R.J., and D.G. Cahill. 1990. Ultrafast time resolution in scanned probe microscopies. *Appl. Phys. Lett.* 57:2031-2033.

(b) Hamers, R.J., and D.G. Cahill. 1991. Ultrafast time resolution in scanned probe microscopies: Surface photovoltage on Si(111)-(7x7). *J. Vac. Sci. Technol. B* 9:514-518.

(c) Steeves, G.M., and M.R. Freeman. 2002. Ultrafast scanning tunneling microscopy. *Adv. Imag. Electron Phys.* 125:195-229.

(d) Nunes, G., and M.R. Freeman. 1993. Picosecond resolution in scanning tunneling microscopy. *Science* 262:1029-1032.

(e) Kelly, K.F., Z.J. Donhauser, B.A. Mantooth, and P.S. Weiss. 2005. Expanding the capabilities of the scanning tunneling microscope. In *Scanning Probe Microscopy: Characterization, Nanofabrication and Device Application of Functional Materials*, Vilarinho P.M., Y. Rosenwaks, and A. Kingon, eds., Kluwer Academic. Vol. 186, 152-171.

147. Hamers, R.J. 1989. Atomic-resolution surface spectroscopy with the scanning tunneling microscopy. *Annu. Rev. Phys. Chem.* 40:531-559.

148. Feenstra, R.M., J.A. Stroscio, J. Tersoff, and A.P. Fein. 1987. Atom-selective imaging of the GaAs(110) surface. *Phys. Rev. Lett.* 58:1192-1197.

149. Sykes, E.C.H., P. Han, S.A. Kandel, K.F. Kelly, G.S. McCarty, and P.S. Weiss. 2003. Substrate-mediated interactions and intermolecular forces between molecules adsorbed on surfaces. *Acc. Chem. Res.* 36:945-953.

150. Berndt, R., and J. K. Gimzewski. 1991. Inelastic tunneling excitation of tip-induced plasmon modes on noble-metal surfaces. *Phys. Rev. Lett.* 67:3796-3799.

151. Alvarado, S.F., P. Renaud, D.L. Abraham, C.H. Schonenberger, D.J. Arent, and H.P. Meier. 1991. Luminescence in scanning tunneling microscopy on III-V nanostructures. *J. Vac. Sci. Technol. B* 9:409-413.

152. Egusa, S., Y.-H. Liau, and N.F. Scherer. 2004. Imaging scanning tunneling microscope-induced electroluminescence in plasmonic corrals. *Appl. Phys. Lett.* 84:1257-1259.

153. Haynes, C.L., and R.P. Van Duyne. 2001. Nanosphere lithography: A versatile nanofabrication tool for studies of size-dependent nanoparticle optics. *J. Phys. Chem. B* 105:5599-5611.

154. Fromm, D.P., A. Sundaramurthy, P.J. Schuck, G.S. Kino, and W.E. Moerner. 2004. Gap-dependent optical coupling of single "bowtie" nanoantennas resonant in the visible. *Nano Lett.* 4:957-961.

155. Stipe, B.C., M.A. Rezaei, and W. Ho. 1998. Single-molecule vibrational spectroscopy and microscopy. *Science* 280:1732-1735.

156. Jaklevic, R.C., and J. Lambe. 1966. Molecular vibration spectra by inelastic electron tunneling. *Phys. Rev. Lett.* 165:1139-1142.

157. Albrecht, T.R., S. Akamine, M.J. Zdeblick, and C.F. Quate. 1990. Microfabrication of integrated scanning tunneling microscope. *J. Vac. Sci. Technol. A* 8:317-318.

158. Bale, M., and R.E. Palmer. 2002. Microfabrication of silicon tip structures for multiple-probe scanning tunneling microscopy. *J. Vac. Sci. Technol. B* 20:364-369.

159. Kaiser, W.J., L.D. Bell, M.H. Hecht, and L.C. Davis. 2001. BEEM and the characterization of buried interfaces. In *Scanning Probe Microscopy and Spectroscopy: Theory, Techniques, and Applications*, Bonnell, D.A., ed. New York: Wiley-VCH.

160. (a) Chen, C.J. 1993. *Introduction to Scanning Tunneling Microscopy*. New York: Oxford University Press.

(b) Wiesendanger, R. 1994. *Scanning Probe Microscopy and Spectroscopy: Methods and Applications*. New York: Cambridge University Press.

(c) Bonnell, D.A. 2001. *Scanning Probe Microscopy and Spectroscopy*. New York: Wiley-VCH.

161. Giessibl, F.J. 2005. Atomic force microscopy's path to atomic resolution. *Mat. Today* 8:32-41.

162. Albrecht, T.R., S. Akamine, T.E. Carver, and C.F. Quate. 1990. Microfabrication of cantilever styli for the atomic force microscope. *J. Vac. Sci. Technol. A* 8:3386-3396.

163. Woolley, A.T., C.L. Cheung, J.H. Hafner, and C.M. Lieber. 2000. Structural biology with carbon nanotube AFM probes. *Chem. Biol.* 7:R193-R204.

164. Nguyen, C.V., R.M.D. Stevens, J. Barber, J. Han, and M. Meyyappan. 2002. Carbon nanotube scanning probe for profiling of deep-ultraviolet and 193 nm photoresist patterns. *Appl. Phys. Lett.* 81:901-903.

165. (a) Giessibl, F.J. 1995. Atomic resolution of the silicon (111)-(7x7) surface by atomic force microscopy. *Science* 68:68-71.

(b) Fukui, K.-I., H. Onishi, and Y. Iwasawa. 1997. Atom-resolved image of the $TiO_2(110)$ surface by noncontact atomic force microscopy. *Phys. Rev. Lett.* 79:4202-4205.

166. Ohnesorge, F., and G. Binnig. 1993. True atomic resolution by atomic force microscopy through repulsive and attractive forces. *Science* 260:1451-1456.

167. (a) Stolz, M., D. Stoffler, U. Aebi, and C. Goldsbury. 2000. Monitoring biomolecular interactions by time-lapse atomic force microscopy. *J. Struct. Biol.* 131:171-180.

(b) Lindsay, S.M. 2001. The scanning probe microscope in biology. *Scanning Probe Microscopy and Spectroscopy: Theory, Techniques, and Applications,* Bonnell, D.A., ed. New York: Wiley-VCH.

168. Horber, J.K.H., and M. J. Miles. 2003. Scanning probe evolution in biology. *Science* 302:1002-1005.

169. Findeis, M.A. 2000. Approaches to discovery and characterization of inhibitors of amyloid beta-peptide polymerization. *Biochim. Biophys. Acta* 1502:76-84.

170. Sulchek, T., G.G.Yaralioglu, C.F. Quate, and S.C. Minne. 2002. Characterization and optimization of scan speed for tapping-mode atomic force microscopy. *Rev. Sci. Instrum.* 73:2928-2936.

171. Karrai, K., and R. D. Grober. 1995. Piezoelectric tip-sample distance control for near-field optical microscopes. *Appl. Phys. Lett.* 66:1842-1844.

172. Sulchek, T., R. Hsieh, J.D. Adams, G G. Yaralioglu, S.C. Minne, C.F. Quate, J.P. Cleveland, A. Atalar, and D. M. Adderton. 2000. High-speed tapping mode imaging with active Q control for atomic force microscopy. *Appl. Phys. Lett.* 76:1473-1475. See also note 159.

173. (a) Bukofsky, S.J., and R.D. Grober. 1997. Video rate near-field scanning optical microscopy. *Appl. Phys. Lett.* 71:2749-2751.

(b) Magnussen, O.M., W. Polewska, L. Zitzler, and R. J. Behm. 2001. In situ video-STM studies of dynamic processes at electrochemical interfaces. *Faraday Discuss.* 121:43-52.

(c) Frenken, J.W.M., T.H. Oosterkamp, B.L.M. Hendriksen, and M.J. Rost. 2005. Pushing the limits of SPM. *Mat. Today* 8:20-25.

(d) Besenbacher, F., E. Laegsgaard, and I. Stensgaard. 2005. Fast scanning STM studies. *Mat. Today* 8:26-30.

(e) Rost, M.J., L. Crama, P. Schakel, E. van Tol, G. B.E.M. van Velzen-Williams, C.F. Overgauw, H. ter Horst, H. Dekker, B. Okhuijsen, M. Seynen, A. Vijftigschild, P. Han, A.J. Katan, K. Schoots, R. Schumm, W. van Loo, T. H. Oosterkamp, and J.W.M. Frenken. 2005. Scanning probe microscopes go video rate and beyond. *Rev. Sci. Instrum.* 76:053710.

174. Humphris, A.D.L., J.K. Hobbs, and M.J. Miles. 2003. Ultrahigh-speed scanning near-field optical microscopy capable of over 100 frames per second. *Appl. Phys. Lett.* 83:6-8.

175. (a) Takano, H., J.R. Kenseth, S.-S. Wong, J.C. O'Brien, and M.D. Porter. 1999. Chemical and biochemical analysis using scanning force microscopy. *Chem. Rev.* 99:2845-2890.

(b) Frisbie, C.D., L.F. Rozsnyai, A. Noy, M.S. Wrighton, and C.M. Lieber. 1994. Functional group imaging by chemical force microscopy. *Science* 265:2071-2073.

(c) Noy, A., D.V. Vezenov, and C.M. Lieber. 1997. Chemical force microscopy. *Annu. Rev. Mater. Sci.* 27:381-421.

176. Israelachvili, J. 1987. Solvation forces and liquid structure, as probed by direct force measurements. *Acc. Chem. Res.* 20:415-421.

177. (a) Carpick, R.W., and M. Salmeron. 1997. Scratching the surface: Fundamental investigations of tribology with atomic force microscopy. *Chem. Rev.* 97:1163-1194.

(b) Takano, H., J.R. Kenseth, S.-S. Wong, J.C. O'Brien, and M.D. Porter. 1999. Chemical and biochemical analysis using scanning force microscopy. *Chem. Rev.* 99: 2845-2890.

(c) Frisbie, C.D., L.F. Rozsnyai, A. Noy, M.S. Wrighton, and C.M. Lieber. 1994. Functional group imaging by chemical force microscopy. *Science* 265:2071-2073.

178. Hoh, J.H., J.P. Cleveland, C.B. Prater, J.-P. Revel, and P.K. Hansma. 1992. Quantized adhesion detected with the atomic force microscope. *J. Am. Chem. Soc.* 114:4917-4918.

179. Frisbie, C.D., L.F. Rozsnyai, A.Noy, M.S. Wrighton, and C.M. Lieber. 1994. Functional group imaging by chemical force microscopy. *Science* 265:2071-2073.

180. (a) Florin, E.-L., V.T. Moy, and H.E. Gaub. 1994. Adhesion forces between individual ligand-receptor pairs. *Science* 264:415-417.

(b) Lee, G.U., L.A. Chrisey, and R.J. Colton. 1994. Direct measurement of the forces between complementary strands of DNA. *Science* 266:771-773.

(c) Boland, T., and B.D. Ratner. 1995. Direct measurement of hydrogen bonding in DNA nucleotide bases by atomic force microscopy. *Proc. Natl. Acad. Sci. USA* 92:5297-5301.

181. (a) Noy, A., D.V. Vezenov, and C.M. Lieber. 1997. Chemical force microscopy. *Annu. Rev. Mater. Sci.* 27:381-421.

(b) Noy, A., C.H. Sanders, D.V. Vezenov, S.S. Wond, and C.M. Lieber. 1998. Chemically sensitive imaging in tapping mode by chemical force microscopy: Relationship between phase lag and adhesion. *Langmuir* 14:1508-1511.

182. Hillborg, H., N. Tomczak, A. Olah, H. Schonherr, and G.J. Vansco. 2004. Nanoscale hydrophobic recovery: A chemical force microscopy study of UV/ozone-treated cross-linked poly(dimethylsiloxane). *Langmuir* 20:785-794.

183. Ibid.

184. Takano, H., J.R. Kenseth, S.-S. Wong, J.C. O'Brien, and M.D. Porter. 1999. Chemical and biochemical analysis using scanning force microscopy. *Chem. Rev.* 99:2845-2890.

185. Ibid.

186. Frisbie, C.D., L.F. Rozsnyai, A. Noy, M.S. Wrighton, and C.M. Lieber. 1994. Functional group imaging by chemical force microscopy. *Science* 265:2071-2073.

187. Florin, E.-L., V.T. Moy, and H.E. Gaub. 1994. Adhesion forces between individual ligand-receptor pairs. *Science* 264:415-417.

188. Boland, T., and B.D. Ratner. 1995. Direct measurement of hydrogen bonding in DNA nucleotide bases by atomic force microscopy. *Proc. Natl. Acad. Sci. USA* 92:5297-5301.

189. (a) Lee, G.U., L.A. Chrisey, and R.J. Colton. 1994. Direct measurement of the forces between complementary strands of DNA. *Science* 266:771-773.

(b) Strunz, T., K. Oroszlan, R. Schafer, and H. -J. Guntherodt. 1999. Dynamic force spectroscopy of single DNA molecules. *Proc. Natl. Acad. Sci. USA* 96:11277-11282.

(c) Noy, A. 2004. Direct determination of the equilibrium unbinding potential profile for a short DNA duplex from force spectroscopy data. *Appl. Phys. Lett.* 85:4792-4794.

(d) Ling, L., H.-J. Butt, and R. Berger. 2004. Rupture force between the third strand and the double strand within a triplex DNA. *J. Am. Chem. Soc.* 126:13992-13997.

190. Williams, J.M., T. Han, and T.P. Beebe. 1996. Determination of single-bond forces from contact force variances in atomic force microscopy. *Langmuir* 12:1291-1295.

191. (a) Fujihira, M. 1999. Kelvin probe force microscopy of molecular surfaces. *Ann. Rev. Mater. Sci.* 29:353-380.

(b) Nonnenmacher, M., M.P. O'Boyle, and H.K. Wickramasinghe. 1991. Kelvin probe force microscopy. *Appl. Phys. Lett.* 58:2921-2923.

192. Fujihira, M. 1999. Kelvin probe force microscopy of molecular surfaces. *Ann. Rev. Mater. Sci.* 29:353-380.

193. Muller, E.M., and J.A. Marohn. 2005. Microscopic evidence for spatially inhomogeneous charge trapping in pentacene. *Adv. Mater.* 17:1410-1414.

194. Johnson, A.S., C.L. Nehl, M.G. Mason, and J.H. Hafner. 2003. Fluid electric force microscopy for charge density mapping in biological systems. *Langmuir* 19:10007-10010.

195. Wilson, N.R., and J.V. MacPherson. 2004. Enhanced resolution electric force microscopy with single-wall carbon nanotube tips. *J. Appl. Phys.* 96:3565-3567.

196. (a) Kramer, S., R.R. Fuierer, and C.B. Gorman. 2003. Scanning probe lithography using self-assembled monolayers. *Chem. Rev.* 103:4367-4418.

(b) Liu, G.-Y., S. Xu, and Y. Qian. 2000. Nanofabrication of self-assembled monolayers using scanning probe lithography. *Acc. Chem. Res.* 33:457-466.

(c) Nyffenegger, R.M., and R.M. Penner. 1997. Nanometer-scale surface modification using the scanning probe microscope: Progress since 1991. *Chem. Rev.* 97:1195-1230.

197. Piner, R.D., J. Zhu, F. Xu, S. Hong, and C.A. Mirkin. 1999. "Dip-pen" nanolithography. *Science* 283:661-663.

198. Liu, G.-Y., S. Xu, and Y. Qian. 2000. Nanofabrication of self-assembled monolayers using scanning probe lithography. *Acc. Chem. Res.* 33:457-466.

199. Schonherr, H., V. Chechik, C.J.M. Stirling, and G.J. Vansco. 2000. Monitoring surface reactions at an AFM tip: An approach to follow reaction kinetics in self-assembled monolayers on the nanometer scale. *J. Am. Chem. Soc.* 122:3679-3687.

200. (a) Minne, S.C., G. Yaralioglu, S.R. Manalis, J.D. Adams, J. Zesch, A. Atalar, and C.F. Quate. 1998. Automated parallel high-speed atomic force microscopy. *Appl. Phys. Lett.* 72:2340-2342.

(b) Saya, D., K. Fukushima, H. Toshiyoshi, G. Hashiguchi, H. Fujita, and H. Kawakatsu. 2002. Fabrication of single-crystal Si cantilever array. *Sens. Actuators A* A95:281-287.

(c) Chow, E.M., G.G. Yaralioglu, C.F. Quate, and T.W. Kenny. 2002. Characterization of a two-dimensional cantilever array with through-wafer electrical interconnects. *Appl. Phys. Lett.* 80:664-666.

201. Piner, R.D., J. Zhu, F. Xu, S. Hong, and C.A. Mirkin. 1999. "Dip-pen" nanolithography. *Science* 283:661-663.

202. Liu, G.-Y., S. Xu, and Y. Qian. 2000. Nanofabrication of self-assembled monolayers using scanning probe lithography. *Acc. Chem. Res.* 33:457-466.

203. Ibid.

204. (a) Higgins, D.A., D.A. Vanden Bout, J. Kerimo, and P.F. Barbara. 1996. Polarization-modulation near field scanning optical microscopy of mesostructured materials. *J. Phys. Chem.* 100:13794-13803.

(b) Higgins, D.A., J. Kerimo, D.A. Vanden Bout, and P.F. Barbara. 1996. A molecular yarn: Near field optical studies of self assembled, flexible, fluorescent fibers. *J. Am. Chem. Soc.* 118:4049-4058.

(c) Higgins, D.A., X. Liao, J. Hall, and E. Mei. 2001. Simultaneous near-field optical birefringence and fluorescence contrast applied to the study of dye-doped polymer-dispersed liquid crystals. *J. Phys. Chem. B* 105:5874-5882.

(d) Vickery, S.A., and R.C. Dunn. 2001. Direct observation of structural evolution in palmitic acid monolayers following Langmuir-Blodgett deposition. *Langmuir* 17:8204-8209.

(e) Aoki, H., and S. Ito. 2001. Two-dimensional polymer investigated by scanning near-field optical microscopy: Phase separation of polymer blend monolayer. *J. Phys. Chem. B* 105:4558-4564.

205. (a) Higgins, D.A., X. Liao, J. Hall, and E. Mei. 2001. Simultaneous near-field optical birefringence and fluorescence contrast applied to the study of dye-doped polymer-dispersed liquid crystals. *J. Phys. Chem. B* 105:5874-5882.

(b) Aoki, H., and S. Ito. 2001. Two-dimensional polymer investigated by scanning near-field optical microscopy: Phase separation of polymer blend monolayer. *J. Phys. Chem. B* 105:4558-4564.

206. Hwang, J., L.K. Tamm, C. Bohm, T.S. Ramalingam, E. Betzig, and M. Edidin. 1995. Nanoscale complexity of phospholipid monolayers investigated by near-field scanning optical microscopy. *Science* 270:610-614.

207. Vickery, S.A., and R.C. Dunn. 2001. Direct observation of structural evolution in palmitic acid monolayers following Langmuir-Blodgett deposition. *Langmuir* 17:8204-8209.

208. (a) Garcia-Parajo, M.F., J.-A. Veerman, R. Bouwhuis, R. Vvallee, and N.F. van Hulst. 2001. Optical probing of single fluorescent molecules and proteins. *ChemPhysChem* 2:347-360.

(b) Höppener, C., J.P. Siebrasse, R. Peters, U. Kubitscheck, and A. Naber. 2005. High-resolution near-field optical imaging of single nuclear pore complexes under physiological conditions. *Biophys. J.* 88:3681-3688.

(c) Enderle, T., T. Ha, D.F. Ogletree, D.S. Chemla, C. Magowan, and S. Weiss. 1997. Membrane specific mapping and colocalization of malarial and host skeletal proteins in the *Plasmodium falciparum* infected erythrocyte by dual-color near-field scanning optical microscopy. *Proc. Natl. Acad. Sci. USA* 94:520-525.

209. (a) Betzig, E., and R.J. Chichester. 1993. Single molecules observed by near-field scanning optical microscopy. *Science* 262:1422-1425.

(b) Ruiter, A.G.T., J.A. Veerman, M.F. Garcia-Parajo, and N.F. van Hulst. 1997. Single molecule rotational and translational diffusion observed by near-field scanning optical microscopy. *J. Phys. Chem. A* 101:7318-7323.

210. Higgins, D.A., J. Kerimo, D.A. Vanden Bout, and P.F. Barbara. 1996. A molecular yarn: Near field optical studies of self assembled, flexible, fluorescent fibers. *J. Am. Chem. Soc.* 118:4049-4058.

211. Teetsov, J.A., and D.A. Vanden Bout. 2001. Imaging molecular and nanoscale order in conjugated polymer thin films with near-field scanning optical microscopy. *J. Am. Chem. Soc.* 123:3605-3606.

212. Dragnea, B., J. Preusser, W. Schade, S.R. Leone, and W.D. Hinsberg. 1999. Transmission near-field scanning microscope for infrared chemical imaging. *J. Appl. Phys.* 86:2795-2799.

213. (a) Akhremitchev, B.B., S. Pollack, and G.C. Walker. 2001. Apertureless scanning near-field infrared microscopy of a rough polymeric surface. *Langmuir* 17:2774-2781.

(b) Lahrech, A., R. Bachelot, P. Gleyzes, and A.C. Boccara. 1997. Infrared near-field imaging of implanted semiconductors: Evidence of a pure dielectric contrast. *Appl. Phys. Lett.* 71:575-577.

(c) Keilmann, F. 2002. Vibrational-infrared near-field microscopy. *Vibrat.Spectrosc.* 29: 109-114.

214. (a) Jahncke, C.L., M.A. Paesler, and H.D. Hallen. 1995. Raman imaging with near-field scanning optical microscopy. *Appl. Phys. Lett.* 67:2483-2485.

(b) Anderson, N., A. Hartschuh, and L. Novotny. 2005. Near-field Raman microscopy. *Mat. Today* 8:50-54.

215. Zeisel, D., B. Dutoit, V. Deckert, T. Roth, and R. Zenobi. 1997. Optical spectroscopy and laser desorption on a nanometer scale. *Anal. Chem.* 69:749-754.

216. (a) Hartschuh, A., E.J. Sanchez, X.S. Xie, and L. Novotny. 2003. High-resolution near-field Raman microscopy of single-walled carbon nanotubes. *Phys. Rev. Lett.* 90:095503.

(b) Pettinger, B., B. Ren, G. Picardi, R. Schuster, and G. Ertl. 2004. Nanoscale probing of adsorbed species by tip-enhanced Raman spectroscopy. *Phys. Rev. Lett.* 92:096101.

217. Jahncke, C.L., M.A. Paesler, and H.D. Hallen. 1995. Raman imaging with near-field scanning optical microscopy. *Appl. Phys. Lett.* 67:2483-2485.

218. (a) Veerman, J.A., A.M. Otter, L. Kuipers, and N.F. van Hulst. 1998. High definition aperture probes for near-field optical microscopy fabricated by focused ion beam milling. *Appl. Phys. Lett.* 72:3115-3117.

(b) Pilevar, S., K. Edinger, W. Atia, I. Smolyaninov, and C. Davis. 1998. Focused ion-beam fabrication of fiber probes with well-defined apertures for use in near-field scanning optical microscopy. *Appl. Phys. Lett.* 72:3133-3135.

219. Betzig, E., and R.J. Chichester. 1993. Single molecules observed by near-field scanning optical microscopy. *Science* 262:1422-1425.

220. Hollars, C.W., and R.C. Dunn. 2000. Probing single molecule orientations in model lipid membranes with near-field scanning optical microscopy. *J. Chem. Phys.* 112:7822-7830.

221. Higgins, D.A., J. Kerimo, D.A. Vanden Bout, and P.F. Barbara. 1996. A molecular yarn: Near field optical studies of self assembled, flexible, fluorescent fibers. *J. Am. Chem. Soc.* 118:4049-4058.

222. (a) Teetsov, J.A., and D.A. Vanden Bout. 2001. Imaging molecular and nanoscale order in conjugated polymer thin films with near-field scanning optical microscopy. *J. Am. Chem. Soc.* 123:3605-3606.

(b) DeAro, J.A., K.D. Weston, S.K. Buratto, and U. Lemmer. 1997. Mesoscale optical properties of conjugated polymers probed by near-field scanning optical microscopy. *Chem. Phys. Lett.* 277:532-538.

(c) Blatchford, J.W., T.L. Gustafson, A.J. Epstein, D.A. Vanden Bout, J. Kerimo, D.A. Higgins, P.F. Barbara, D.-K. Fu, T.M. Swager, and A.G. MacDiarmid. 1996. Spatially and temporally resolved emission from aggregates in conjugated polymers. *Phys. Rev. B* 54:R3683-R3686.

223. Paesler, M.A., and P.J. Moyer. 1996. *Near-Field Optics: Theory, Instrumentation, and Applications.* New York: John Wiley & Sons.

224. (a) Higgins, D.A., D.A. Vanden Bout, J. Kerimo, and P.F. Barbara. 1996. Polarization-modulation near field scanning optical microscopy of mesostructured materials. *J. Phys. Chem.* 100:13794-13803.

(b) Vaez-Iravani, M., and R. Toledo-Crow. 1993. Pure linear polarization imaging in near field scanning optical microscopy. *Appl. Phys. Lett.* 63:138-140.

(c) McDaniel, E.B., S.C. McClain, and J.W.P. Hsu. 1998. Nanometer scale polarimetry studies using a near-field scanning optical microscope. *Appl. Opt.* 37:84-92.

(d) Goldner, L.S., M.J. Fasolka, S. Noufler, H.-P. Nguyen, G.W. Bryant, J. Hwang, K.D. Weston, K.L. Beers, A. Urbas, and E.L. Thomas. 2003. Fourier analysis near-field polarimetry for measurement of local optical properties of thin films. *Appl. Opt.* 42:3864-3881.

(e) Lacoste, T., T. Huser, R. Prioli, and H. Heinzelmann. 1998. Contrast enhancement using polarization-modulation scanning near-field optical microscopy. *Ultramicroscopy* 71:333-340.

225. Humphris, A.D.L., J.K. Hobbs, and M.J. Miles. 2003. Ultrahigh-speed scanning near-field optical microscopy capable of over 100 frames per second. *Appl. Phys. Lett.* 83:6-8.

226. Bukofsky, S.J., and R.D. Grober. 1997. Video rate near-field scanning optical microscopy. *Appl. Phys. Lett.* 71:2749-2751.

227. (a) McNeill, J.D., and P.F. Barbara. 2002. NSOM investigation of carrier generation, recombination, and drift in a conjugated polymer. *J. Phys. Chem. B* 106:4632-4639.

(b) Higgins, D.A., J.E. Hall, and A. Xie. 2005. Optical microscopy studies of dynamic within individual polymer-dispersed liquid crystal droplets. *Acc. Chem. Res.* 38:137-145.

228. (a) Levy, J., V. Nikitin, J.M. Kikkawa, D.D. Awschalom, and N. Samarth. 1996. Femtosecond near-field spin microscopy in digital magnetic heterostructures. *J. Appl. Phys.* 97:6095-6100.

(b) Guenther, T., C. Lienau, T. Elsaesser, M. Glanemann, V.M. Axt, T. Kuhn, S. Eshlaghi, and A.D. Wieck. 2002. Coherent nonlinear optical response of single quantum dots studied by ultrafast near-field spectroscopy. *Phys. Rev. Lett.* 89:057401.

229. (a) McNeill, J.D., and P.F. Barbara. 2002. NSOM investigation of carrier generation, recombination, and drift in a conjugated polymer. *J. Phys. Chem. B* 106:4632-4639.

(b) Adams, D.M., J. Kerimo, C.-Y. Liu, A.J. Bard, and P.F. Barbara. 2000. Electric field modulated near-field photo-luminescence of organic thin films. *J. Phys. Chem. B* 104:6728-6736.

230. Higgins, D.A., J.E. Hall, and A. Xie. 2005. Optical microscopy studies of dynamic within individual polymer-dispersed liquid crystal droplets. *Acc. Chem. Res.* 38:137-145.

231. Davy, S., and M. Spajer. 1996. Near field optics: Snapshot of the field emitted by a nanosource using a photosensitive polymer. *Appl. Phys. Lett.* 69:3306-3308.

232. (a) H'dhili, F., R. Bachelot, G. Lerondel, D. Barchiesi, and P. Royer. 2001. Near-field optics:

Direct observation of the field enhancement below and apertureless probe using a photosensitive polymer. *Appl. Phys. Lett.* 79:4019-4021.

(b) Tarun, A.., M.R.H. Daza, N. Hayazawa, Y. Inouye, and S. Kawata. 2002. Apertureless optical near-field fabrication using an atomic force microscope on photoresists. *Appl. Phys. Lett.* 80:3400-3402.

(c) Yin, X., N. Fang, X. Zhang, I.B. Martini, and B.J. Schwartz. 2002. Near-field two-photon nanolithography using an apertureless optical probe. *Appl. Phys. Lett.* 81:3663-3665.

233. Yin, X., N. Fang, X. Zhang, I.B. Martini, and B.J. Schwartz. 2002. Near-field two-photon nanolithography using an apertureless optical probe. *Appl. Phys. Lett.* 81:3663-3665.

234. Davy, S., and M. Spajer. 1996. Near field optics: Snapshot of the field emitted by a nanosource using a photosensitive polymer. *Appl. Phys. Lett.* 69:3306-3308.

235. (a) Tarun, A.., M.R.H. Daza, N. Hayazawa, Y. Inouye, and S. Kawata. 2002. Apertureless optical near-field fabrication using an atomic force microscope on photoresists. *Appl. Phys. Lett.* 80:3400-3402.

(b) Yin, X., N. Fang, X. Zhang, I.B. Martini, and B.J. Schwartz. 2002. Near-field two-photon nanolithography using an apertureless optical probe. *Appl. Phys. Lett.* 81:3663-3665.

236. Sun, S., and G.J. Leggett. 2004. Matching the resolution of electron beam lithography by scanning near-field photolithography. *Nano Lett.* 4:1381-1384.

237. Lee, M.B., N. Atoda, K. Tsutsui, and M. Ohtsu. 1999. Nanometric aperture arrays fabricated by wet and dry etching of silicon for near-field optical storage application. *J. Vac. Sci. Technol. B* 17:2462-2466.

238. Bukofsky, S.J., and R.D. Grober. 1997. Video rate near-field scanning optical microscopy. *Appl. Phys. Lett.* 71:2749-2751.

239. Sanchez, E.J., L. Novotny, and X.S. Xie. 1999. Near-field fluorescence microscopy based on two-photon excitation with metal tips. *Phys. Rev. Lett.* 82:4014-4017.

240. (a) Jahncke, C.L., M.A. Paesler, and H.D. Hallen. 1995. Raman imaging with near-field scanning optical microscopy. *Appl. Phys. Lett.* 67:2483-2485.

(b) Anderson, N., A. Hartschuh, and L. Novotny. 2005. Near-field Raman microscopy. *Mat. Today* 8:50-54.

(c) Hartschuh, A., E.J. Sanchez, X.S. Xie, and L. Novotny. 2003. High-resolution near-field Raman microscopy of single-walled carbon nanotubes. *Phys. Rev. Lett.* 90:095503.

(d) Pettinger, B., B. Ren, G. Picardi, R. Schuster, and G. Ertl. 2004. Nanoscale probing of adsorbed species by tip-enhanced Raman spectroscopy. *Phys. Rev. Lett.* 92:096101.

241. Krug, J.T., E.J. Sanchez, and X.S. Xie. 2002. Design of near-field optical probes with optimal field enhancement by finite difference time domain electromagnetic simulation. *J. Chem. Phys.* 116:10895-10901.

242. (a) Novotny, L., E.J. Sanchez, and X.S. Xie. 1998. Near-field optical imaging using metal tips illuminated by higher-order Hermite-Gaussian beams. *Ultramicroscopy* 71:21-29.

(b) Sanchez, E.J., L. Novotny, and X.S. Xie. 1999. Near-field fluorescence microscopy based on two-photon excitation with metal tips. *Phys. Rev. Lett.* 82:4014-4017.

243. (a) Adams, D.M., J. Kerimo, C.-Y. Liu, A.J. Bard, and P.F. Barbara. 2000. Electric field modulated near-field photo-luminescence of organic thin films. *J. Phys. Chem. B* 104:6728-6736.

(b) Mei, E., and D.A. Higgins. 1998. Near-field optical microscopy studies of electric-field-induced molecular reorientation dynamics. *J. Phys. Chem. A* 102:7558-7563.

244. (a) Adams, D.M., J. Kerimo, C.-Y. Liu, A.J. Bard, and P.F. Barbara. 2000. Electric field modulated near-field photo-luminescence of organic thin films. *J. Phys. Chem. B* 104:6728-6736.

(b) Mei, E., and D.A. Higgins. 1998. Near-field optical microscopy studies of electric-field-induced molecular reorientation dynamics. *J. Phys. Chem. A* 102:7558-7563.

(c) Moerner, W.E., T. Plakhotnik, T. Irngartinger, U.P. Wild, D.W. Pohl, and B. Hecht. 1994. Near-field optical spectroscopy of individual molecules in solids. *Phys. Rev. Lett.* 73:2764-2767.

245. (a) DeAro, J.A., D. Moses, and S.K. Buratto. 1999. Near-field photoconductivity of stretch-oriented poly(para-phenylene vinylene). *Appl. Phys. Lett.* 75:3814-3816.

(b) Gray, M.H., J.W.P. Hsu, L. Giovane, and M.T. Bulsara. 2001. Effect of anisotropic strain on the crosshatch electrical activity in relaxed GeSi films. *Phys. Rev. Lett.* 86:3598-3601.

(c) McNeill, C.R., H. Frohne, J.L. Holdworth, and P.C. Dastoor. 2004. Near-field scanning photocurrent measurements of polyfluorene blend devices: Directly correlating morphology with current generation. *Nano Lett.* 4:2503-2507.

246. Kwak, J., and A.J. Bard. 1989. Scanning electrochemical microscopy. Theory of the feedback mode. *Anal. Chem.* 61:1221-1227.

247. (a) James, P.J., L.F. Garfias-Mesias, P.J. Moyer, and W.H. Smyrl. 1998. Scanning electrochemical microscopy with simultaneous independent topography. *J. Electrochem. Soc.* 145:L64-L66.

(b) Lee, Y., Z. Ding, and A.J. Bard. 2002. Combined scanning electrochemical/optical microscopy with shear force and current feedback. *Anal. Chem.* 74:3634-3643.

248. Ibid.

249. Fan, F.-R.F., and A.J. Bard. 1999. Imaging of biological macromolecules on mica in humid air by scanning electrochemical microscopy. *Proc. Nat. Acad. Sci. USA* 96:14222-14227.

250. Basame, S.B., and H.S. White. 1999. Scanning electrochemical microscopy of metal/metal oxide electrodes. Analysis of spatially localized electron-transfer reactions during oxide growth. *Anal. Chem.* 71:3166-3170.

251. Fernandez, J.L., D.A. Walsh, and A.J. Bard. 2005. Thermodynamic guidelines for the design of bimetallic catalysts for oxygen electroreduction and rapid screening by scanning electrochemical microscopy. M-Co (M: Pd, Ag, Au). *J. Am. Chem. Soc.* 127:357-365.

252. Liu, B., S.A. Rotenberg, and M.V. Mirkin. 2000. Scanning electrochemical microscopy of living cells: Different redox activities of nonmetastatic and metastatic human breast cells. *Proc. Nat. Acad. Sci. USA* 97:9855-9860.

253. Liu, B., A.J. Bard, M.V. Mirkin, and S.E. Creager. 2004. Electron transfer at self-assembled monolayers measured by scanning electrochemical microscopy. *J. Am. Chem. Soc.* 126:1485-1492.

254. Zhang, J., C.J. Slevin, C. Morton, P. Scott, D.J. Walton, and P.J. Unwin. 2001. New approach for measuring lateral diffusion in Langmuir monolayers by scanning electrochemical microscopy (SECM): Theory and application. *J. Phys. Chem. B* 105:11120-11130.

255. (a) Zhang, J., C.J. Slevin, C. Morton, P. Scott, D.J. Walton, and P.J. Unwin. 2001. New approach for measuring lateral diffusion in Langmuir monolayers by scanning electrochemical microscopy (SECM): Theory and application. *J. Phys. Chem. B* 105:11120-11130.

(b) O'Mullane, A.P., J.V. Macpherson, P.R.Unwin, J. Cervera-Montesinos, J.A. Manzanares, F. Frehill, J.G. Vos. 2004. Measurement of lateral charge propagation in [Os(Bpy)2(Pvp)Ncl]Cl thin films: A scanning electrochemical microscopy approach. *J. Phys. Chem. B* 108:7219-7227.

256. (a) Bath, B.D., H.S. White, and E.R. Scott. 2000. Electrically facilitated molecular transport. Analysis of the relative contributions of diffusion, migration, and electroosmosis to solute transport in an ion-exchange membrane. *Anal. Chem.* 72:433-442.

(b) Uitto, O.D., and H.S. White. 2001. Scanning electrochemical microscopy of membrane transport in the reverse imaging mode. *Anal. Chem.* 72:533-539.

257. (a) Liu, B., S.A. Rotenberg, and M.V. Mirkin. 2000. Scanning electrochemical microscopy of living cells: Different redox activities of nonmetastatic and metastatic human breast cells. *Proc. Nat. Acad. Sci. USA* 97:9855-9860.

(b) Yasykawa, T., T. Kaya, and T. Matsue. 1999. Dual imaging of topography and photo-synthetic activity of a single protoplast by scanning electrochemical microscopy. *Anal. Chem.* 71:4637-4641.

258. Liu, B., S.A. Rotenberg, and M.V. Mirkin. 2000. Scanning electrochemical microscopy of living cells: Different redox activities of nonmetastatic and metastatic human breast cells. *Proc. Nat. Acad. Sci. USA* 97:9855-9860.

259. Wittstock, G., and W. Schuhmann. 1999. Formation and imaging of microscopic enzymatically

active spots on an alkanethiolate-covered gold electrode by scanning electrochemical microscopy. *Anal. Chem.* 69:5059-5066.

260. Turyan, I., T. Matsue, and D. Mandler. 2000. Patterning and characterization of surfaces with organic and biological molecules by the scanning electrochemical microscope. *Anal. Chem.* 72:3431-3435.

261. Frenken, J.W.M., T.H. Oosterkamp, B.L.M. Hendriksen, and M.J. Rost. 2005. Pushing the limits of SPM. *Mat. Today* 8:20-25.

262. (a) James, P.J., L.F. Garfias-Mesias, P.J. Moyer, and W.H. Smyrl. 1998. Scanning electrochemical microscopy with simultaneous independent topography. *J. Electrochem. Soc.* 145:L64-L66.

(b) Lee, Y., Z. Ding, A.J. Bard. 2002. Combined scanning electrochemical/optical microscopy with shear force and current feedback. *Anal. Chem.* 74:3634-3643.

263. Gray, M.H., J.W.P. Hsu, L. Giovane, M.T. Bulsara. 2001. Effect of anisotropic strain on the crosshatch electrical activity in relaxed GeSi films. *Phys. Rev. Lett.* 86:3598-3601.

264. Treutler, T.H., and G. Wittstock. 2003. Combination of an electrochemical tunneling microscope (ECSTM) and a scanning electrochemical microscope (SECM): Application for tip-induced modification of self-assembled monolayers. *Electrochim. Acta* 48:2923-2932.

265. (a) Macpherson, J.V., and P.J. Unwin. 2000. Combined scanning electrochemical-atomic force microscopy. *Anal. Chem.* 72:276-285.

(b) Kranz, C., G. Friedbacher, and B. Mizaikoff. 2001. Integrating an ultramicroelectrode in an AFM cantilever—Combined technology for enhanced information. *Anal. Chem.* 73:2491-2500.

266. (a) James, P.J., L.F. Garfias-Mesias, P.J. Moyer, and W.H. Smyrl. 1998. Scanning electrochemical microscopy with simultaneous independent topography. *J. Electrochem. Soc.* 145:L64-L66.

(b) Lee, Y., Z. Ding, and A.J. Bard. 2002. Combined scanning electrochemical/optical microscopy with shear force and current feedback. *Anal. Chem.* 74:3634-3643.

267. Lee, Y., Z. Ding, A.J. Bard. 2002. Combined scanning electrochemical/optical microscopy with shear force and current feedback. *Anal. Chem.* 74:3634-3643.

268. Alpuche-Aviles, M.A., and D.O. Wipf. 2001. Impedance feedback control for scanning electrochemical microscopy. *Anal. Chem.* 73:4873-4881.

269. (a) Sidles, J.A. 1992. Folded Stern-Gerlach experiment as a means for detecting nuclear magnetic resonance in individual nuclei. *Phys. Rev. Lett.* 68:1124-1127.

(b) Rugar, D., C.S. Yannoni, and J.A. Sidles. 1992. Mechanical detection of"5agnetic resonance. *Nature* 360:563-566.

270. (a) Gremlich, H.U., and B. Yan. 2001. *Infrared and Raman Spectroscopy of Biological Materials.* New York: Marcel Dekker.

(b) Lewis, E.N., P.J. Treado, R.C. Reeder, G.M. Story, A.E. Dowres, C. Marcott, and I.W. Levin. 1995. FTIR spectroscopic imaging using an infrared focal-plane array detector. *Anal. Chem.* 67:3377-3381.

(c) Kotula, P.G., M.R. Keenan, and J.R. Michael. 2003. Automated analysis of SEM X-ray spectral images: A powerful new microanalysis tool. *Microsc. Microanal.* 9:1-17.

(d) Artyushkova, K., and J. Fulghum. 2003. XPS imaging. In *Surface Analysis by Auger and X-ray Photoelectron Spectroscopy*, D. Briggs, and J.T. Grant, eds. Chichester, Manchester, U.K.: IM publications and SurfaceSpectra Limited.

(e) Winograd, N. 2003. Prospects or imaging TOF-SIMS: From fundamentals to biotechnology. *Appl. Surf. Sci.* 203:13-19.

(f) Treado, P.J., M.P. Nelson, and T.H. Myers. 2000. Current state of the art in Raman microscopy and chemical imaging. *Microbeam Anal. 2000,Proc.* 165:37-38.

271. (a) Oberholzer, M., M. Ostreicher, H. Christen, and M. Bruhlmann. 1996. Methods in quantitative image analysis. *Histochem. Cell Biol.* 105:333-355.

(b) Bovik, A. 2000. *Handbook of Image and Video Processing.* San Diego, CA: Academic Press.

272. Wilson, D.L., K.S. Kump, S.J. Eppell, and R.E. Marchant. 1995. Morphological restoration of atomic-force microscopy images. *Langmuir* 11:265-272.

273. Jak, M.J.J., C. Konstapel, A. van Kreuningen, J. Verhoeven, R. van Gastel, and J.W.M. Frenken. 2001. Automated detection of particles, clusters and islands in scanning probe microscopy images. *Surf. Sci.* 494:43-52.

274. Meijering, E.H.W., W.J. Niessen, and M.A. Viergever. 2001. Quantitative evaluation of convolution-based methods for medical image interpolation. *Med. Image Anal.* 5:111-126.

275. (a) Henderson, R. 1995. The potential and limitations of neutrons, electrons and X-rays for atomic resolution microscopy of unstained biological molecules. *Q. Rev. Biophys.* 28(2):171-193.

(b) Bajaj, C., Z.Y. Yu, and M. Auer. 2003. Volumetric feature extraction and visualization of tomographic molecular imaging. *J. Struct. Biol.* 144:132-143.

(c) Baldock, B., C.J. Gilpin, A.J. Koster, U. Ziese, K.E. Kadler, C.M. Kielty, and D.F. Holmes. 2002. Three-dimensional reconstructions of extracellular matrix polymers using automated electron tomography. *J. Struct. Biol.* 138:130-136.

276. Al-Kofahi, K.A., A. Can, S. Lasek, D.H. Szarowski, N. Dowell-Mesfin, W. Shain, J.T. Turner, and B. Roysam. 2003. Median-based robust algorithms for tracing neurons from noisy confocal microscope images. *IEEE Trans. Info. Tech. Biomed.* 7:302-317.

277. Udupa, J.K. 1999. Three-dimensional visualization and analysis methodologies: A current perspective. *Radiographics* 19:783-806.

278. (a) Leong, F.J.W.M., M. Brady, and J.O. McGee. 2003. Correction of uneven illumination (vignetting) in digital microscopy images. *J.Clin. Pathol.y* 56:619-621.

(b) Tomazevic, D., B. Likar, and F. Pernus. 2002. Comparative evaluation of retrospective shading correction methods. *J. Microsc.* 208:212-223.

279. (a) Lippert, T., E. Ortelli, J.C. Panitz, F. Raimondli, J. Wambach, J. Wei, and A. Wokuan. 1999. Imaging—XPS/Raman investigation on the carbonization of polyimide after irradiation at 308 nm. *Appl. Phys. A.* 69:S651-S654.

(b) Artyushkova, K., and J.E. Fulghum. 2004. Multivariate analysis applied to topographical corrections in X-ray photoelectron images. *Surf. Inter. Anal.* 36:1304-1313.

(c) Allen, G.C., K.R. Hallam, J.R. Eastman, G.J. Graveling, V.K. Ragnarsdottir, and D.R. Skuse. 1998. XPS analysis of polyacrylamide adsorption to kaolinite, quartz and feldspar. *Surf. Interface Anal.* 26:518-523.

280. (a) Jak, M.J.J., C. Konstapel, A. van Kreuningen, J. Verhoeven, R. van Gastel, and J.W.M. Frenken. 2001. Automated detection of particles, clusters and islands in scanning probe microscopy images. *Surf. Sci.* 494:43-52.

(b) Lippert, T., E. Ortelli, J.C. Panitz, F. Raimondli, J. Wambach, J. Wei, and A. Wokuan. 1999. Imaging—XPS /Raman investigation on the carbonization of polyimide after irradiation at 308 nm. *Appl. Phys. A.* 69:S651-S654.

281. (a) Oberholzer, M ., M. Ostreicher, H. Christen, and M. Bruhlmann. 1996. Methods in quantitative image analysis. *Histochem. Cell Biol.* 105:333-355.

(b) Tyler, B. 2003. Interpretation of TOF-SIMS images: Multivariate and univariate approaches to image de-noising, image segmentation and compound identification. *Appl. Surf. Sci.* 203-204:825-831.

(c) Newsam, S., S. Bhagavathy, C. Kenney, B.S. Manjunath, and L. Fonseca. 2001. Object-based representations of spatial images. *Acta Astronaut.* 48:567-577.

282. Oberholzer, M ., M. Ostreicher, H. Christen, and M. Bruhlmann. 1996. Methods in quantitative image analysis. *Histochem.Cell Biol.* 105:333-355.

283. Wilson, D.L., K.S. Kump, S.J. Eppell, and R.E. Marchant. 1995. Morphological restoration of atomic-force microscopy images. *Langmuir* 11:265-272.

284. (a) Newsam, S., S. Bhagavathy, C. Kenney, B.S. Manjunath, and L. Fonseca. 2001. Object-based representations of spatial images. *Acta Astronaut.* 48:567-577.

(b) Bernasconi, A., S.B. Antel, D.L. Collins, N. Bernasconi, A. Olivier, F. Dubeau, G.B. Pike,

F. Andermann, and D.L. Arnold. 2001. Texture analysis and morphological processing of magnetic resonance imaging assist detection of focal cortical dysplasia in extra-temporal partial epilepsy. *Ann. Neurol.* 49:770-775.

285. (a) Dur, J.C., F. Elsass, V. Chaplain, and D. Tessier. 2004. The relationship between particle-size distribution by laser granulometry and image analysis by transmission electron microscopy in a soil clay fraction. *European J. Soil Sci.*55:265-270.

(b) Costa, L.D., C.A. Rodrigues, N.C. de Souza, and O.N. Oliveira. 2003. Statistical characterization of morphological features of layer-by-layer polymer films by image analysis. *J. Nanosci. Nanotech.* 3:257-261.

286. (a) Rasa, M., and A.P. Philipse. 2002. Scanning probe microscopy on magnetic colloidal particles. *J. Magn. Magn. Mat.* 252:101-103.

(b) Plaschke, M., T. Schafer, T. Bundschuh, T.N. Manh, R. Knopp, H. Geckeis, and J.I. Kim. 2001. Size characterization of bentonite colloids by different methods. *Anal. Chem.* 73:4338-4347.

287. (a) Gerlich, D., J. Mattes, and R. Eils. 2003. Quantitative motion analysis and visualization of cellular structures. *Methods* 29:3-13.

(b) Sbalzarini, I.F., and P. Koumoutsakos. 2005. Feature point tracking and trajectory analysis for video imaging in cell biology. *J. Struct. Biol.* 151:182-195.

288. (a) Ashin, R., A. Morimoto, M. Nagase, and R. Vaillancourt. 2005. Image compression with multiresolution singular value decomposition and other methods. *Math. Comp. Modeling* 41:773-790.

(b) Schelkens, P., A. Munteanu, J. Barbarien, M. Galca, X. Giro-Nieto, and J. Cornelis. 2003. Wavelet coding of volumetric medical datasets. *IEEE Trans. Medi. Imag.* 22:441-458.

289. Wickes, B.T., Y. Kim, and D.G. Castner. 2003. Denoising and multivariate analysis of time-of-flight SIMS images. *Surface and Interface Analysis* 35:640-648.

290. (a) Daubechies, I. 1988. Orthnormal bases of compactly supported wavelets. *Commu. Pure Appl.Math.* 41:909-996.

(b) Mallat, S. 1989. A theory for multiresolution signal decomposition. *IEEE Tran. Pattern Anal. Machine Intell.* 11:674-693.

291. Starck, J.L., and A. Bijaoui. 1994. A signal processing filtering and deconvolution by the wavelet transform. *Signal Proc.* 35:195-211.

292. Wolkenstein, M., H. Hutter, S.G. Nikolov, M. Grasserbauer. 1997. Improvement of SIMS image classification by means of wavelet de-noising. *Fresenius J. Anal.l Chem.* 357:783-788.

293. (a) Barequet, G., and M. Sharir. 1996. Piecewise-linear interpolation between polygonal slices. *Comput. Vis. Image Und.* 63:251-272.

(b) Goshtasby, A., D.A. Turner, and L.V. Ackerman. 1992. Matching of tomographic image slices. *IEEE Trans. Med. Imaging* 11:507-516.

(c) Chen, E., and L. Williams. 1993. View interpolation for image synthesis. SIGGRAPH '93, *http://citeseer.ist.psu.edu/chen93view.html.*

294. (a) Geladi, P., and G. Hans. 1996. *Multivariate Image Analysis.* New York: John Wiley & Sons.

(b) Geladi, P., H. Isaksson, L. Lindquist, S. Wold, and K. Esbensen. 1989. Principal component analysis of multivariate images. *Chemomet. Intell. Lab. Sys.* 5:209-220.

(c) de Juan, A., R. Tauler, R. Dyson, C. Marcolli, M. Rault, and M. Maeder. 2004. Spectroscopic imaging and chemometrics: A powerful combination for global and local sample analysis. *Trac-Trends Anal. Chem.* 23:70-79.

(d) Richards, J.A. 1999. *Remote Sensing Digital Image Analysis: An Introduction.* New York: Springer.

(e) Stubbings, T.C., M.G. Wolkenstein, and H. Hutter. 1999. Comparison of different approaches to analytical images classification. *J. Trace Microprobe Tech.* 17:1-16.

(f) Prutton, M., D.K. Wilkinson, P.G. Kenny, and D.L. Mountain. 1999. Data processing for spectrum-images: Extracting information from the data mountain. *Appl. Surf. Sci.* 145:1-10.

295. (a) Geladi, P., H. Isaksson, L. Lindquist, S. Wold, K. Esbensen. 1989. Principal component analysis of multivariate images. *Chemomet. Intell. Lab. Sys.* 5:209-220.

(b) Prutton, M., D.K. Wilkinson, P.G. Kenny, and D.L. Mountain. 1999. Data processing for spectrum-images: Extracting information from the data mountain. *Appl. Surf. Sci.* 145:1-10.

296. (a) Wickes, B.T., Y. Kim, and D.G. Castner. 2003. Denoising and multivariate analysis of time-of-flight SIMS images. *Surf.Interface Anal.*35:640-648.

(b) Coullerez, G., S. Lundmark, E. Malmstrom, A. Hult, and H.J. Mathieu. 2003. ToF-SIMS for the characterization of hyperbranched aliphatic polyesters: Probing their molecular weight on surfaces based on principal component analysis (PCA). *Surf. Interface Anal.* 35:693-708.

(c) Peterson, R.E., and B.J. Tyler. 2003. Surface composition of atmospheric aerosol: Individual particle characterization by TOF-SIMS. *Appl. Surf. Sci.* 203:751-756.

(d) Tyler, B. 2003. Interpretation of TOF-SIMS images: Multivariate and univariate approaches to image de-noising, image segmentation and compound identification. *Appl. Surf. Sci.* 203:825-831.

297. (a) Kotula, P.G., M.R. Keenan, and J.R. Michael. 2003. Automated analysis of SEM X-ray spectral images: A powerful new microanalysis tool. *Microsc. Microanal.*9:1-17.

(b) Roseman, A.M. 2004. FindEM—A fast, efficient program for automatic selection of particles from electron micrographs. *J. Struct. Biol.* 145:91-99.

(c) Jembrih, D., M. Schreiner, M. Peev, P. Krejsa, and C. Clausen. 2000. Identification and classification of iridescent glass artifacts with XRF and SEM/EDX. *Mikrochim Acta* 133:151-157.

(d) Flesche, H., A.A. Nielsen, and R. Larsen. 2000. Supervised mineral classification with semiautomatic training and validation set generation in scanning electron microscope energy dispersive spectroscopy images of thin sections. *Math. Geol.* 32:337-366.

(e) Egelandsdal, B., K.F. Christiansen, V. Host, F. Lundby, J.P. Wold, and K. Kvaal. 1999. Evaluation of scanning electron microscopy images of a model dressing using image feature extraction techniques and principal component analysis. *Scanning* 21:316-325.

298. (a) Artyushkova, K., and J.E. Fulghum. 2002. Multivariate image analysis methods applied to XPS imaging data sets. *Surf. Interface Anal.* 33:185-195.

(b) Peebles, D.E., J.A. Ohlhausen, P.G. Kotula, S. Hutton, and C. Blomfield. 2004. Multivariate statistical analysis for x-ray photoelectron spectroscopy spectral imaging: Effect of image acquisition time. *J. Vacuum Sci. Techn. A* 22:1579-1586.

(c) Walton, J., and N. Fairley. 2004. Quantitative surface chemical-state microscopy by X-ray photoelectron spectroscopy. *Surf. Inter. Anal.* 36:89-91.

299. (a) Duponchel, L., W. Elmi-Rayaleh, C. Ruckebusch, and J.P. Huvenne. 2003. Multivariate curve resolution methods in imaging spectroscopy: Influence of extraction methods and instrumental perturbations. *J.Chem. Info. Com. Sci.* 43:2057-2067.

(b) Budevska, B.O., S.T. Sum, and T.J. Jones. 2003. Application of multivariate curve resolution for analysis of FT-IR microspectroscopic images of in situ plant tissue. *Appl. Spectrosc.* 57: 124-131.

(c) Budevska, B.O. 2000. Minimization of optical non-linearities in Fourier transform-infrared microspectroscopic imaging. *Vibrational Spectrosc.* 24:37-45.

300. (a) Zhang, D.M., and D. Ben-Amotz. 2000. Enhanced chemical classification of Raman images in the presence of strong fluorescence interference. *Appl. Spectrosc.* 54:1379-1383.

(b) Sasic, S., D.A. Clark, J.C. Mitchell, and M.J. Snowden. 2004. A comparison of Raman chemical images produced by univariate and multivariate data processing–A simulation with an example from pharmaceutical practice. *Analyst* 129:1001-1007.

(c) Drumm, C.A., and M.D. Morris. 1995. Microscopic Raman line-imaging with principal component analysis. *Appl.Spectrosc.* 49:1331-1337.

(d) Gallagher, N.B., J.M. Shaver, E.B. Martin, J. Morris, B.M. Wise, and W. Windig. 2004. Curve resolution for multivariate images with applications to TOF-SIMS and Raman. *Chemomet. Intell. Lab. Sys.* 73:105-117.

301. (a) Bonnet, N. 1998. Multivariate statistical methods for the analysis of microscope image series: Applications in materials science. *J.Microsc.—Oxford* 190:2-18.

(b) Hadjiiski, L., S. Munster, E. Oesterschulze, and R. Kassing. 1996. Neural network correction of nonlinearities in scanning probe microscope images. *J. Vacuum Sci. Tech. B* 14:1563-1568.

(c) Velthuizen, R.P., L.O. Hall, and L.P. Clarke. 1999. Feature extraction for MRI segmentation. *J. Neuro.* 9:85-90.

302. (a) Roseman, A.M. 2004. FindEM—a fast, efficient program for automatic selection of particles from electron micrographs. *J. Struct. Biol.* 145:91-99.

(b) Jembrih, D., M. Schreiner, M. Peev, P. Krejsa, and C. Clausen. 2000. Identification and classification of iridescent glass artifacts with XRF and SEM/EDX. *Mikrochim. Acta* 133:151-157.

(c) Flesche, H., A.A. Nielsen, and R. Larsen. 2000. Supervised mineral classification with semiautomatic training and validation set generation in scanning electron microscope energy dispersive spectroscopy images of thin sections. *Math. Geol.* 32:337-366.

(d) Egelandsdal, B., K.F. Christiansen, V. Host, F. Lundby, J.P. Wold, and K. Kvaal. 1999. Evaluation of scanning electron microscopy images of a model dressing using image feature extraction techniques and principal component analysis. *Scanning* 21:316-325.

303. (a) Tauler, R., A. Smilde, and B.J. Kowalski. 1995. Selectivity, local rank, 3-way data analysis and ambiguity in multivariate curve resolution. *J. Chemom.* 9:31-58.

(b) Duponchel, L., W. Elmi-Rayaleh, C. Ruckebusch, and J.P. Huvenne. 2003. Multivariate curve resolution methods in imaging spectroscopy: Influence of extraction methods and instrumental perturbations. *J. Chem. Infor. Com. Sci.* 43:2057-2067.

304. (a) Hadjiiski, L., S. Munster, E. Oesterschulze, and R. Kassing. 1996. Neural network correction of nonlinearities in scanning probe microscope images. *J. Vacuum Sci. Tech. B* 14:1563-1568.

(b) Duponchel, L., W. Elmi-Rayaleh, C. Ruckebusch, and J.P. Huvenne. 2003. Multivariate curve resolution methods in imaging spectroscopy: Influence of extraction methods and instrumental perturbations. *J. Chem. Info. Com. Sci.* 43:2057-2067.

305. MATLAB: The Language of Technical Computing. The Mathworks, Inc., Natick, Massachusetts.

306. PLS_toolbox. Eigenvector Research Inc., Wenatchee, Washington.

307. *http://www.mathworks.com/matlabcentral/fileexchange/loadCategory.do.*

308. ENVI: The Environment for Visualizing Images. The Research Systems, Inc., Boulder, Colorado.

309. Kotula, P.G., M.R. Keenan, and J.R. Michael. 2003. Automated analysis of SEM X-ray spectral images: A powerful new microanalysis tool. *Microsc. Microanal.* 9:1-17.

310. Ibid.

311. (a) Smentkowski, V.S., J.A. Ohlhausen, P.G. Kotula, and M.R. Keenan. 2004. Multivariate statistical analysis of time-of-flight secondary ion mass spectrometry images-looking beyond the obvious. *Appl. Surf. Sci.* 231-232:245-249.

(b) Ohlhausen, J.A.T., M.R. Keenan, P.G. Kotula, and D.E. Peebles. 2004. Multivariate statistical analysis of time-of-flight secondary ion mass spectrometry images using AXSIA. *Appl. Surf. Sci.* 231-232:230-234.

312. Peebles, D.E., J.A. Ohlhausen, P.G. Kotula, S. Hutton, and C. Blomfield. 2004. Multivariate statistical analysis for x-ray photoelectron spectroscopy spectral imaging: Effect of image acquisition time. *J. Vacuum Sci. Tech. A* 22:1579-1586.

313. (a) Pohl, C. 1999. Tools and methods for fusion of images of different spatial resolution. *Int.Arch.Photogram. Remote Sens.* 32:W6.

(b) Hall, D.L., and S.A.H. McMullen. 2004. *Mathematical Techniques in Multisensor Data Fusion.* Norwood, Massachusetts: Artech House.

314. Glasbey, C.A., and N.J. Martin. 1996. Multimodal microscopy by digital image processing. *J. Microsc.—Oxford* 181:225-237.

315. Leonardi, A.J., B.A. Blakistone, and S.W. Kyryk. 1990. Application of microscopy in the paper industry: Case histories of the Mead corporation. *Food Struc.* 9:203-213.

316. Stephens, D.J., and V.J. Allan. 2003. Light microscopy techniques for live cell imaging. *Science* 300:82-86.

317. Richards, J.A. 1999. *Remote Sensing Digital Image Analysis: An Introduction.* New York: Springer.

318. (a) Artyushkova, K., and J. Fulghum. 2003. XPS imaging. In *Surface Analysis by Auger and X-ray Photoelectron Spectroscopy*, D. Briggs, and J.T. Grant, eds. Chichester, Manchester, U.K.: IM publications and SurfaceSpectra Limited.

    (b) Artyushkova, K., B. Wall, J. Koenig, and J.E. Fulghum. 2000. Correlative spectroscopic imaging: XPS and FT-IR studies of PVC/PMMA polymer blends. *Appl. Spectrosc.* 54:1549-1558.

319. Clarke, F.C., M.J. Jamieson, D.A. Clark, S.V. Hammond, R.D. Jee, and A.C. Moffat. 2001. Chemical image fusion. The synergy of FT-NIR and Raman mapping microscopy to enable a more complete visualization of pharmaceutical formulations. *Anal. Chem.* 73:2213-2220.

320. Carugo, O., and S. Pongor. 2002. The evolution of structural databases. *Trends Biotechnol.* 20:498-501.

321. (a) Deshpande, N., K.J. Addess, W.F. Bluhm, J.C. Merino-Ott, W. Townsend-Merino, Q. Zhang, C. Knezevich, L. Xie, L. Chen, Z.K. Feng, R.K. Green, J.L. Flippen-Anderson, J. Westbrook, H.M. Berman, and P.E. Bourne. 2005. The RCSB Protein Data Bank: A redesigned query system and relational database based on the mmCIF schema. *Nucleic Acids Res.* 33:D233-D237.

    (b) Venclovas, C., K. Ginalski, and C. Kang. 2004. Sequence-structure mapping errors in the PDB: OB-fold domains. *Protein Sci.* 13:1594-1602.

    (c) Berman, H.M., J. Westbrook, Z. Feng, G. Gilliland, T.N. Bhat, H. Weissig, I.N. Shindyalov, and P.E. Bourne. 2000. The Protein Data Bank. *Nucleic Acids Res.* 28:235-242.

322. (a) Fischer, D., J. Pas, and L. Rychlewski. 2004. The PDB-Preview database: A repository of in-silico models of "on-hold" PDB entries. *Bioinformatics (Oxford)* 20:2482-2484.

    (b) Fitzkee, N.C., P.J. Fleming, and G.D. Rose. 2005. The protein coil library: A structural database of nonhelix, nonstrand fragments derived from the PDB. *Proteins—Struc. Function Bioinform.* 58:852-854.

323. Nattkemper, T.W. 2004. Multivariate image analysis in biomedicine. *J. Biomed. Inform.* 37:380-391.

324. Brown, J., D. Buckley, A. Coulthard, A.K. Dixon, J.M. Dixon, D.F. Easton, R.A. Eeles, D.G. Evans, F.G. Gilbert, M. Graves, C. Hayes, J.P. Jenkins, A.P. Jones, S.F. Keevil, M.O. Leach, G.P. Liney, S.M. Moss, A.R. Padhani, G.J. Parker, L.J. Pointon, B.A. Ponder, T.W. Redpath, J.P. Sloane, L.W. Turnbull, L.G. Walker, and R.M. Warren. 2000. Magnetic resonance imaging screening in women at genetic risk of breast cancer: Imaging and analysis protocol for the UK multicentre study. *Magn. Reson. Imaging* 18(7):765-776.

325. (a) Van Horn, J.D., J.S. Grethe, P. Kostelec, J.B. Woodward, J.A. Aslam, D. Rus, D. Rockmore, and M.S. Gazzaniga. 2001. The Functional Magnetic Resonance Imaging Data Center (fMRIDC): The challenges and rewards of large-scale databasing of neuroimaging studies. *Philosoph. Trans. R. Soc. Lond. B Biol. Sci.* 356:1323-1339.

    (b) Van Horn, J.D., and M.S. Gazzaniga. 2002. Databasing fMRI studies—Towards a "discovery science" of brain function. *Nat. Rev. Neurosci.* 3:314-318.

326. Van Horn, J.D., and M.S. Gazzaniga. 2002. Databasing fMRI studies—Towards a "discovery science" of brain function. *Nat. Rev. Neurosci.* 3:314-318.

327. Gonzalez-Couto, E., B. Hayes, and A. Danckaert. 2001. The life sciences Global Image Database (GID). *Nucleic Acids Res.* 29:336-339.

328. (a) Fayyad, U., D. Haussler, and P. Stolorz. 1996. Mining scientific data. *Commun. ACM* 39:51-57.

    (b) Hsu, W., M.L. Lee, and K.G. Goh. 2000. Image mining in IRIS: Integrated retinal information system. *SIGMOD Record* 29:593-593.

329. Cooke, C.D., C. Ordonez, E.V. Garcia, E. Omiecinski, E.G. Krawczynska, R.D. Folks, C.A. Santana, L. DeBraal, and N.F. Ezquerra. 1999. Data mining of large myocardial perfusion SPECT (MPS) databases to improve diagnostic decision making. *J. Nucl. Med.* 40(suppl.S): 293P-293P.

330. Andrienko, G., and N. Andrienko. 1999. Knowledge-based visualization to support spatial data mining. *Lect. Notes Comp. Sci.* 1642:149-160.

331. (a) Allen, M.P., and D.J. Tildesley. 1989. *Computer Simulation of Liquids.* Oxford: Clarendon Press.

(b) Frenkel, D., and B. Smit. 2002. *Understanding Molecular Simulation.* San Diego, CA: Academic Press.

332. Larson, S.M., C.D. Snow, M. Shirts, and V.S. Pande. 2002. *Computational Genomics,* Grant, R., ed. Norwich, UK: Horizon Press.

333. Todorov, I.T., and W. Smith. 2004. DL_POLY_3: The CCP5 national UK code for molecular-dynamics simulations. One contribution of 12 to a theme "Discrete element modelling: Methods and applications in the environmental sciences." *Phil. Trans. R. Soc. Lond. A* 362:1835-1852.

334. Kendall, R.A., E. Apra, D.E. Bernholdt, E.J. Bylaska, M. Dupuis, G.I. Fann, R.J. Harrison, J. Ju, J.A. Nichols, J. Nieplocha, T.P. Straatsma, T.L. Windus, and A.T. Wong. 2000. High performance computational chemistry: An overview of NWChem a distributed parallel application. *Computer Phys. Comm.* 128:260-283.

335. (a) Brooks, B.R., R.E. Bruccoleri, B.D. Olafson, D.J. States, S. Swaminathan, and M. Karplus. 1983. CHARMM: A program for macromolecular energy, minimization, and dynamics calculations. *J. Comp. Chem.* 4:187-217.

(b) MacKerell, Jr., A.D., B. Brooks, C.L. Brooks, III, L. Nilsson, B. Roux, Y. Won, and M. Karplus. 1998. CHARMM: The energy function and its parameterization with an overview of the program. *The Encyclopedia of Computational Chemistry,* Schleyer, P.v.R., et al., eds. Chichester, U.K.: John Wiley & Sons.

336. (a) Ponder, J.W., and F.M. Richards. 1987. An efficient newton-like method for molecular mechanics energy minimization of large molecules. *J. Comput. Chem.* 8:1016-1024.

(b) Ren, P., and J.W. Ponder. 2002. Consistent treatment of inter- and intramolecular polarization in molecular mechanics calculations. *J. Comput. Chem.* 23:1497-1506.

337. van der Spoel, D., E. Lindahl, B. Hess, A.R. van Buuren, E. Apol, P.J. Meulenhoff, D.P. Tieleman, A.L.T.M. Sijbers, K.A. Feenstra, R. van Drunen, and H.J.C. Berendsen. 2004. *Gromacs User Manual Version 3.2 www.gromacs.org.*

338. Laxmikant, K., R. Skeel, M. Bhandarkar, R. Brunner, A. Gursoy, N. Krawetz, J. Phillips, A. Shinozaki, K. Varadarajan, and K. Schulten. 1999. NAMD2: Greater scalability for parallel molecular dynamics. *J. Comp. Phys.* 151:283-312.

339. Siepmann, J.I. 2001. Challenges in the development of transferable force fields for phase equilibrium calculations. in *Forum 2000: Fluid Properties for New Technologies, Connecting Virtual Design with Physical Reality,* Rainwater, J.C., et al., eds. Pp. 110-112. Boulder, CO: NIST Special Publication.

340. Bolton, K., W.L. Hase, and G.H. Peshlherbe. 1998. *Modern Methods for Multidimensional Dynamics Computation in Chemistry,* Thompson, D.L., ed. Singapore: World Scientific.

341. Car, R., and M. Parrinello. 1985. Unified approach for molecular dynamics and density-functional theory. *Phys. Rev. Lett.* 55:2471.

342. (a) Schlegel, H.B., J.M. Millam, S.S. Iyengar, G.A. Voth, A.D. Daniels, G.E. Scuseria, and M.J. Frisch. 2001. Ab initio molecular dynamics: Propagating the density matrix with Gaussian orbitals. *J. Chem. Phys.* 114:9758-9763.

(b) Iyengar, S.S., H.B. Schlegel, J.M. Millam, G.A. Voth, G.E. Scuseria, and M.J. Frisch. 2001. Ab initio molecular dynamics: Propagating the density matrix with Gaussian orbitals. II. Generalizations based on mass-weighting, idempotency, energy conservation and choice of initial conditions. *J. Chem. Phys.* 115:10291-10302.

(c) Schlegel, H.B., S.S. Iyengar, X. Li, J.M. Millam, G.A. Voth, G.E. Scuseria, and M.J. Frisch. 2002. Ab initio molecular dynamics: Propagating the density matrix with Gaussian orbitals. III. Comparison with Born-Oppenheimer dynamics. *J. Chem. Phys.* 117:8694-8704.

343. (a) Yip, S. 2003. Synergistic materials science. *Nat. Mat.* 2:3.

(b) Yip, S., ed. 2005. *Handbook of Materials Modeling.* New York: Springer.

344. Ibid.

4

Committee Findings and Recommendations

In this report, chemical imaging is defined as the spatial and temporal characterization of the molecular composition, structure, and dynamics of any given sample—with the ultimate goal being able to both *understand* and *control* complex chemical processes. As illustrated by the case studies in Chapter 2, this ability to image or visualize chemical events in space and time is essential to the future development of many fields of science.

At present, imaging lies at the heart of the many advances taking place in our high-technology world. For example, microscopic imaging experiments have played a key role in the development of organic material devices used in electronics. Chemical imaging is also critical to understanding diseases such as Alzheimer's, where chemical imaging provides the ability to determine molecular structure, cell structure, and communication and integrate these into obtaining information nondestructively from the human brain. Continued advances in these chemical imaging capabilities will result in more fundamental understanding of chemical processes. Concurrently, advances in other areas of research—such as nano-science and materials—underlie the developments needed to push chemical imaging ahead even further.

Chemical imaging techniques span a broad array of capabilities and applications. The findings and recommendations described below are offered as guidance for setting priorities and mapping plans toward fundamental breakthroughs in areas of imaging research as well other areas that impact development of chemical imaging. Recommendations are presented in the order in which areas of chemical imaging research were examined in Chapters 2 and 3.

A GRAND CHALLENGE FOR CHEMICAL IMAGING

A very important goal for chemical imaging is to understand and control complex chemical processes. This ultimately requires the ability to perform multimodal or multitechnique imaging across all length and time scales. Complete characterization of a complex material requires information not only on the surface or in bulk chemical components, but also on stereometric features such as size, distance, and homogeneity in three-dimensional space. In chemical imaging, it is frequently difficult to uniquely distinguish between alternative surface morphologies using a single analytical method and routine data acquisition and analysis. Multitechnique image correlation allows for extending lateral and vertical spatial characterization of chemical phases. This approach improves spatial resolution by utilizing techniques with nanometer resolution to enhance data from techniques with micrometer resolution—such as atomic force microscopy (AFM) or scanning electron microscopy (SEM) combined with X-ray photoemission spectroscopy (XPS) or Fourier transform infrared (FTIR) spectroscopy. Multimodal imaging also facilitates correlation of different physical properties such as phase information in AFM with chemical information in XPS. By combining techniques that use different physical principles and record different properties of the object space, complementary and better-quality information becomes available.

As in most cases of systems integration, multimodal imaging requires more than simply networking different imaging techniques. Advances in computational capabilities, for example, are fundamental to effective integration of imaging techniques. Data fusion is the name for the techniques used to combine data from multiple techniques to perform inferences that may not be possible from a single technique by itself. The goal is to combine image data to form a new image that contains more interpretable information than could be gained using the original information. Combining images to form a multimodal image requires—beyond the usual image processing for a single image—a compensation for changes in image alignment from one instrument to another due to slight movements of the specimens, slight differences in magnification, or imperfect centering of the sample. **There is a need to develop multitechnique correlations for various combinations of imaging techniques.**

Two examples for techniques that may be combined are listed below, but they are by no means meant to be comprehensive.

• *Combining Surface Enhanced Raman Spectroscopy and Nanoscale Scanning Probe Techniques*

Surface enhanced Raman spectroscopy (SERS) experiments on silver and gold nanoclusters have demonstrated large enhancement levels and field confinement of 5 nm or less for various samples such as single-walled carbon nanotubes.[1] However, the locations of these conditions cannot be controlled but are instead determined by the specific nanostructures used. That is, the target molecules have to be in the close vicinity of SERS-active nanometer-sized silver or gold substrates. On the other hand, location can be controlled in so-called tip-enhanced SERS experiments.[2] Unfortunately, these experiments provide only small SERS enhancement factors compared with samples interacting with metal nanoparticles. Further development of tip-enhanced technology would benefit from experimental systems that combine high SERS enhancement factors and highly confined probed volumes with nanoscale-controlled scanning. This may be accomplished by combining nanoscale scanning probe techniques (such as modified atomic force microscopy [AFM] systems) with the techniques of single-molecule Raman spectroscopy.

- *Combining X-rays, Electrons, and Scanning Probe Microscopies*

    Scanning probe microscopy (SPM) techniques typically provide topographical, not chemical identification, so that combining other local spectroscopies with STM is typically necessary to identify the atoms and molecules present. Specialized approaches are being developed to address this. For example, progress has been made in joining transmission electron microscopy (TEM) and scanning electron microscopy (SEM) to scanning tunneling microscopy (STM). STM photoemission spectroscopy (PESTM) combined with inelastic electron tunneling spectroscopy yields vibrational and other information. Integrating the three techniques will enable investigation of the chemical (X-ray, infrared, or Raman), structural (EM), and topographic (SPM) nature of samples.

AREAS OF IMAGING RESEARCH

Understanding and controlling complex chemical processes also requires advances in more focused areas of imaging research. These chemical imaging techniques span a broad array of capabilities and applications, and are discussed in great detail within this report. Here, we briefly highlight the research and development that will best advance current capabilities—with a focus on applications in which investment would most likely lead to proportionally large returns. The main findings of the committee are:

Nuclear Magnetic Resonance

Nuclear magnetic resonance (NMR) and magnetic resonance imaging (MRI) represent mature technologies that have widespread impact on the materials, chemical, biochemical, and medical fields. NMR and MRI are very useful tools for obtaining structural and spatial information. It is clear that in the coming years, NMR and MRI will continue to expand rapidly and continue to be key tools for chemical imaging. However, the major limiting factor for application of these techniques to a broader range of problems is their relatively low sensitivity, which is a result of the low radio-frequency energy used. There are a number of ways that need to be explored to obtain more signals from NMR and MRI, and these are provided below:

Detector Signal

A major limiting factor of NMR and MRI is the relatively low sensitivity of their detectors. Progress has been made in increasing detector sensitivity by using supercooled detectors (increasing sensitivity by a factor of 2 to 4), and further progress could be made with new materials both for higher-temperature super-conductors and for better insulation. Work with other detector strategies such as superconducting quantum interference devices (SQUID) and other novel magne-tometers should be encouraged, especially in light of progress in hyperpolarization (see below). Finally, force detection of magnetic resonance is a very promising area that is limited by detector design.

For MRI, gains in sensitivity by a factor of two- to fivefold have been real-ized primarily by building parallel arrays of MRI detectors. Cross-talk between the array detectors limits performance, and the configuration of these detectors is bulky. Novel approaches to building MRI array detectors will make dense arrays possible. Once the construction of dense arrays using small coils is achieved, the noise in the coils themselves will become limiting to signal to noise. It should be possible to cool the coils (as has been done in high-resolution NMR). Making cold, dense, parallel arrays feasible should enable at least a tenfold increase in sensitivity for both NMR and MRI. This will increase the resolution that can be obtained with MRI as well as with MR spectroscopic imaging of a large number of metabolites. **Increasing signal-to-noise ratios should be a chief focus of the efforts to improve the sensitivity of NMR and MRI detectors.**

Hyperpolarization

Another very promising avenue for increasing sensitivity in NMR and MRI is to increase signal from the molecules being detected. In recent years there has been growing interest in hyperpolarization techniques, which couple the nuclear spins being detected by NMR to other spins with a higher polarization, that increase NMR sensitivity by factors of 100-100,000. Exciting developments for hyperpolarization using a variety of techniques—such as dynamic nuclear polar-ization, laser-induced hyperpolarization of noble gases, and formation of parahydrogen—have realized extraordinary gains in sensitivity for applications to materials research and biomedical imaging. At present only a restricted set of molecules has been hyperpolarized using only a small set of possible techniques. **There is a need to expand the range of techniques useful for hyperpolarizing NMR and MRI signals, as well as the range of molecules that can be hyper-polarized.**

Contrast Agents

Strategies for new MRI contrast agents parallel the development of fluorescence probes; however, MRI contrast agents are more typically used in animal models and humans. Therefore, more emphasis is placed on the safety of these contrast agents than on their usefulness as probes for wider uses. It is critical that the MRI relaxivity of contrast agents be improved so that they can be used in smaller quantities. In addition, much progress has been achieved over the past five years in the development of MRI probes that are site-specific as well as capable of tracing particular biological and chemical processes. Most of this has been proof-of-principle work; the more difficult task of optimizing these approaches to ensure their robustness must now be undertaken. Finally, preliminary work in identifying MRI-active proteins or protein assemblies that are equivalent to fluorescent proteins has begun. This is an area in which great gains can be made. **MRI probes need to have higher relaxivity, be more specific, and be deliverable to the site of action.**

Magnet Size

The sensitivity of magnetic resonance also increases with higher magnetic fields. In the range where detector noise dominates, sensitivity increases as approximately the square of the increase in field. In practice, this is difficult to realize, particularly because many samples of interest contribute to noise, leading to an increase in sensitivity that is linearly proportional to magnetic field strength. Nonetheless, much interest has been focused on producing higher magnetic fields, which means larger magnets and larger (expensive) dedicated facilities to house them. There is work now being done to decrease the siting requirement of high-field magnets, for example by employing innovative designs for superconducting wire that can carry higher current densities. These efforts could decrease the size of magnets, enabling very-high-field NMR and MRI to transition from dedicated laboratories to widespread use for applications such as advanced oil exploration, homeland security, and environmental study. **The miniaturization of high-field NMR and MRI magnets is needed to broaden the applicability of these techniques by reducing the need for dedicated facilities.**

Optical Imaging

In contrast to NMR and MRI, optical spectroscopy imaging techniques utilize radiation at an energy level high enough to allow individual photons to be measured relatively easily with modern equipment at a detection sensitivity almost matched by the mammalian eye. As a result, imaging data are acquired at the sensitivity of individual molecules. The inherent temporal and spatial resolution is also increased proportionately, but the resonance itself is broad because environmental influences are not averaged out within the inherent time scale of inter-action between the molecules and this frequency of radiation. As a result, the chemical structural information content of optical spectra is considerably lower than that of magnetic resonance, particularly in the electronic region of the spectrum. Thus, research needs in the area of optical imaging are focused more on increasing chemical structural information.

New Probes Based on Metallic Particles

In terms of the high content of chemical structural information at desired spatial and temporal resolutions, Raman spectroscopy has the potential to be a very useful technique for chemical imaging. However, a disadvantage in many applications of Raman imaging results from relatively poor signal-to-noise ratios due to the extremely small cross section of the Raman process, 12 to 14 orders of magnitude lower than fluorescence cross sections. New methodologies such as localized SERS utilizing metallic nanoparticles can be used to overcome this shortcoming. Metallic nanoparticles have long been used in Mie scattering dark-field microscopy. However, nonspherical particles and their aggregates, which cannot be described by classic Mie scattering theory, offer rich optical properties associated with surface plasma-related phenomena. In recent years, this field has experienced much research activity due to the ability to fabricate new nanostruc-tures, the emergence of sensitive microscopes and detectors, and the availability of tools for electromagnetism computation. However, these advances represent only the beginning of this research area; much work is still needed. For example, almost 30 years after its discovery, there exists little quantitative or even qualita-tive understanding of SERS, which arises in part from a strongly enhanced elec-tric field in the close vicinity of gold and silver nanostructures. Better under-standing of radiation signals—including Raman scattering, Mie scattering, and fluorescence—from the nanostructures or atomic clusters is an important prereq-uisite for the creation of new optical probes. In particular, probes that exploit SERS signals show promise in providing specific spectroscopic signatures and multiplex capabilities along with chemical specificity.

There is a need to develop a better theoretical understanding of the radiation signals of gold and silver nanostructures including Raman scattering, Mie scattering, and fluorescence. New probes composed of metal-based nano-particles or atomic clusters should be developed to provide improved sensi-tivity, specificity, and spatial localization capabilities.

Fluorescent Labels for Bioimaging

Unlike NMR spectroscopy and vibrational spectroscopy, electronic spectroscopy involves interactions with electromagnetic waves in the near-infrared, visible, and ultraviolet (UV) spectral regions. While electronic spectroscopy is less enlightening about structural information than NMR and vibrational spectroscopy, the shorter wavelengths involved allow higher spatial resolution for imaging, and its stronger signal yields superb sensitivity. Fluorescence detection, with its background-free measurement, is especially sensitive and makes single fluorescent molecules detectable. On the other hand, particularly under ambient conditions, the amount of molecular structural information that can be obtained from fluorescence imaging is limited.

Organic fluorophores or labels that bind specifically to macromolecules, metabolites, and ions provide powerful tools for chemical imaging in cells and tissues. For example, green fluorescent protein and its derivatives allow live cell imaging and tracking of individual proteins. In addition, techniques such as fluorescence correlation spectroscopy (FCS), fluorescence resonance energy transfer (FRET), and multiphoton microscopy may be used for localization studies as well as for some cases of chemical reaction dynamics research. However, the efficiencies of chemical and biological labels are hampered by photobleaching. There is a great need to develop more robust labels. To accomplish this, one must understand the photophysics and photochemistry of fluorescent labels as well as the mechanisms of photobleaching. Suggested substitutes, such as semiconducting nanoparticle "quantum dots," have been limited by their intermittent blinking and by the large size required for water-resistant coatings. Promising routes for increased label robustness include dye-molecule clusters fixed in silica shells and photochemically switchable labels.

Besides needing to be sufficiently robust, bright, nontoxic, and small in size, labels must also demonstrate chemical or biological specificity, which is of key importance. Further development of fluorescent labels for widespread application of dynamics research would also benefit the broader chemical imaging community. For example, in biological contexts the incorporation of unnatural fluorescent amino acids into nascent polypeptide chains by genetic encoding is a promising approach.

In order to probe chemical constituents and follow their biochemical reaction in cells and tissues, there is a need to make fluorescent labels more specific, brighter, and more robust. This will require greater understanding of the photophysics and photochemistry of fluorescent probes and the mechanisms of their photobleaching.

Nonlinear Optical Techniques

In addition to imaging based on single-photon excited or linear Raman scattering, vibrational images can also be generated using nonlinear coherent Raman spectroscopies. The most prominent nonlinear Raman process for imaging is coherent anti-Stokes Raman scattering (CARS). Like spontaneous Raman microscopy, CARS microscopy does not rely on natural or artificial fluorescent labels, thereby avoiding issues of toxicity and artifacts associated with staining and photobleaching of fluorophores. Instead, it depends on a chemical contrast intrinsic to the samples. CARS microscopy offers two distinct advantages over conventional Raman microscopy: (1) The radiation damage is significantly less for CARS than for spontaneous Raman, especially when one is interested in following a dynamic process with short data collection time; and (2) it has three-dimensional sectioning capability because the nonlinear CARS signal is generated only at the laser focus where laser intensities are highest. This is particularly useful for imaging thick tissues or cell structures. Techniques such as CARS microscopy and other nonlinear Raman methods offer the possibility of new contrast mechanisms with chemical sensitivity, but their potential depends critically on advances in laser sources, detection schemes, and new Raman labels. Efforts have been made to circumvent the diffraction limit by engineering the point spread function using nonlinear optical techniques. Continued developments in these nonlinear approaches will enable superhigh resolution using far-field optics without the need to employ proximal probes. Multiphoton fluorescence microscopy can also benefit from the development of more compact ultrafast lasers, fiber delivery, and improved fluorophores with larger nonlinear polarizability.

Nonlinear optical techniques need to be developed—with particular emphasis on improved ultrafast laser sources and special fluorophores, novel contrast mechanisms based on nonlinear methods for breaking the diffraction barrier without using proximal probes.

Ultrafast Optical Detectors

Current streak camera technologies allow one to measure the lifetime and spectral features of fluorescence with subpicosecond and subnanometer resolution, but they lack the sensitivity required for single-molecule applications. Charge-coupled device (CCD) cameras, on the other hand, can provide high spectral and/or spatial resolution at high quantum efficiency, but they lack temporal resolution and near-infrared (IR) sensitivity. Improvements in time resolution would also be a boon to lifetime imaging and single-molecule experiments using photon counting avalanche photodiodes, especially with the extension of the spectral range of the detector to near-IR (NIR) and UV regions. IR sensitivity, as has been pointed out, is an especially critical area for improvement, because IR and UV detectors have lagged behind visible detectors in most respects; yet IR provides some of the richest information about chemical structure. Although such advances are likely to be incremental, the effects of sustained successive improvements would be transformative. As yet, the tremendous power of combining modern ultrafast laser technology with high spectral resolution in spatially imaged measurements at nanoscale spatial resolution has not been realized, but this kind of multidimensional measurement is precisely what is required to follow the dynamics of complex interacting mixtures of chemical species.

There is a need for detectors to be developed that possess *all* of the following

attributes: (1) the ability to measure multiple dimensions in parallel fashion, (2) high time resolution, (3) high sensitivity, and (4) broad spectral range. IR and UV detector improvements, even if incremental, could catalyze new chemical insights.

Electron and X-ray Imaging

Techniques that probe samples with wavelengths much smaller than that of visible light provide high-resolution chemical and structural information below surfaces of materials.   With wavelengths that are about 1,000 times smaller than that of visible light, electrons provide a high-resolution probe of chemical and structural information below surfaces of materials. Images of atomic arrangements over a large range of length scales can be obtained using EM techniques. X-rays can penetrate materials much more deeply than either visible light or electrons, producing chemical images that cannot be obtained by any other means.

Sources for Electron Microscopy

A limiting factor in EM is the quality of the electron beam used to probe the sample.   Aberrations introduced by the optics limit both spatial resolution and analytical capabilities. Correcting for spherical and chromatic aberrations introduced by the electron optics will directly improve resolution and other analytical techniques.   Imaging and diffraction will be directly improved by the greater coherence in the beam. In particular, advances in imaging techniques will permit the analysis of amorphous samples. Smaller beam sizes can be achieved, allowing for sub-Angstrom resolution chemical analysis of samples. In addition, aberration correction will relax the constraint on the sample volume, allowing for "a lab-in-the-microscope" approach to in situ microscopy. This will open a vast range of imaging environments such as reactivity measurements, mechanical deformation, and materials synthesis processes.
There is a need to develop higher-quality electron beams in order to broaden and deepen the application of electron microscopy.

Electron Microscopy Detectors

Improved detectors are also needed for EM to enable higher time resolution for imaging chemical kinetics. Higher sensitivity detectors will reduce the amount of electrons needed to image, therefore minimizing the damage that may occur from the electron beam. This will greatly expand the in situ environments for EM that are vital for imaging chemistry, such as reaction dynamics and electron-sensitive materials, including organics and biological samples. Such capabilities are likely to be achieved by developing high-density detector arrays coupled to fast discrimination electronics.

Spatial and temporal sensitivity of electron microscopy detectors need to be improved.

Optics for X-ray Microscopy

Zone plates are diffractive optics for X-ray microscopy that use constructive interference of light rays from adjacent zones to focus. Present zone plates are extremely inefficient (10-20 percent). As a result, a choice must often be made between (1) high efficiency with minimum radiation dose using lower-resolution optics and (2) the highest possible resolution. For trace element mapping in microprobes or mapping nanoscale chemical heterogeneities in spectromicroscopy, higher spatial resolution translates into the ability to carry out chemical imaging at a finer scale with less biasing of quantification. In addition, for immunolabeling, the label size must be comparable to the zone plate resolution in order to be detected; the development of higher-resolution optics will allow the use of smaller labels, which are dramatically easier to coax across the membrane of a cell. Efforts to improve resolution and efficiency require substantial work in nanofabrication in order to make zone plates that push the resolution frontier. Development of optics to correct spherical and chromatic aberrations will greatly improve the resolution of photoemission electron microscopes (PEEM) at third-generation synchrotrons. The development of ultrafast X-ray sources will expand the capabilities of X-ray imaging in terms of space and time.

There is a need to improve zone plate optics, which are presently the limiting factor for scanning transmission X-ray microscopes (STXM) and full-field X-ray microscopes (TXM).

X-ray Detectors

The most common type of X-ray detector is the CCD. In its current incarnation, X-rays from the sample are imaged on a phosphor screen. This converts the X-rays to light, which is then transferred to the CCD chip by means of fiber optics or lens systems. Eliminating this conversion step, and imaging the X-rays directly onto a CCD chip with column parallel readout, will result in a detector with significantly greater sensitivity, higher resolution, and about a hundredfold faster readout speed than today's generation of detectors. These devices would complement the current developments in novel, sophisticated, soft X-ray techniques, such as full-field, deconvolution tomography, which depends on the ability to collect large numbers of high-resolution images rapidly.

Great promise also lies in the further development of solid-state "pixel detectors" for X-ray imaging. Small versions of this detector type are now being used in several industrial and medical imaging applications. However, the technology is relatively immature and requires significant improvement to make it suitable for use in advanced research applications. Another important need is the development of detectors for hard X-ray tomography. The primary focus should be in the area of scintillators for the conversion of X-rays to visible light. These scintillators have to be analyzed at high spatial resolution by light microscope techniques in order to reduce the pixel size of the X-ray to light conversion materials. Finally, an important new class of detectors for chemical imaging is the ultralow-temperature, high energy resolution, solid-state, energy-dispersive detector. A major effort is also required in the area of large-area, low-temperature detector arrays. The astronomy community has already made significant progress in this area, and it is timely to apply these impressive advances to chemical imaging. This has particular application to X-ray microscope-based fluorescence imaging and X-ray spectral imaging of cells and organometallic protein systems. **X-ray detectors—including solid-state pixel detectors and detectors for hard X-ray tomography—need to be improved through the development of scintillators that convert X-rays to visible light and detectors that image directly onto a CCD chip with column parallel readout, among other detector possibilities. The goal is to improve X-ray detector's resolution, dynamic range, sensitivity, and readout speed.**

Probes for X-ray Imaging

The use of X-ray microscopy to image chemical signals in biological materials requires probes that can be applied to both endogenous (naturally occurring) and exogenous (artificially introduced) molecules, particularly proteins. In most cases, endogenous proteins are identified using well-established immunochemistry techniques. However, application of these techniques to X-ray microscopy requires the conjugation of X-ray-dense moieties (that either absorb or become excited by X-rays) to the antigenic molecule. To date, gold-conjugated antibodies have been used to image a single species of protein inside a cell. The challenge now is to develop the capability to simultaneously detect *multiple* proteins inside the cell. This necessitates the development of multiple probes, each of which would contain a specific metal atom that could be excited at a different X-ray energy (e.g., nanocrystals containing atoms such as Ti, V, Fe, and Ni). Chemical signals from specific cellular structures can also be visualized, giving rise to high-resolution information about multiple proteins and interaction partnerships in context. Furthermore, based on the X-ray absorption coefficient, all measurements can be quantified in terms of concentration as well as location. On the other hand, the detection of specific chemical signals from exogenous molecules in the cell requires the development of probes analogous to green fluorescent protein (GFP) that also possess X-ray-absorbing powers. In both cases, these specific probes will facilitate three-dimensional localization of chemical signals at an isotropic resolution approaching 15 nm from whole, hydrated cells. **There is a need to advance X-ray-absorbing probes to specifically detect and localize chemical signals that are introduced into cells.**

Proximal Probes

Since the advent of the scanning tunneling microscope in the early 1980s, a wide variety of related microscopies using similar experimental principles and instrumentation have been developed for imaging samples based on their electronic, optical, chemical, mechanical, and magnetic properties. All find broad applications in high-resolution chemical imaging experiments. Proximal probe microscopes employ a variety of materials—or probes—such as tungsten wire (STM), silicon nitride pyramid and cantilever (AFM), or optical fiber (near-field optical microscopy) in close proximity to the sample of interest for the purposes of recording an image of the sample, performing spectroscopic experiments, or manipulating the sample. These methods are especially useful for understanding the chemistry of surfaces—for example, the electrophilicity of individual surface atoms, the organization of atoms or molecules at or near the surface, and the electronic properties of atomic or molecular assemblies. Two areas of research that would help expand these capabilities are highlighted below.

Penetration Depth

Most high-resolution imaging techniques in materials science are limited to imaging surfaces or near-surface regions. Imaging below surfaces would allow studies of chemistry at the atomic or molecular level occurring at buried interfaces and/or defects sites in the bulk of samples. For example, the development of proximal probe methods (e.g., magnetic resonance force microscopy) by which images of samples can be recorded with high spatial resolution in all three dimensions would represent a major breakthrough in chemical imaging technology. At present, this method is limited primarily by the need for more sensitive cantilevers and stronger magnetic field gradients. Advances in these areas (e.g., force measurement tools for general applications) would allow more sensitive detection and greater spatial resolution to be achieved.

There is a need to develop methods for optical, X-ray, Raman, and other probe regimes that can image at depths of a few nanometers to macroscopic distances beneath a surface, especially for materials science applications.

Chemical Selectivity

Many interesting materials systems are chemically heterogeneous on a wide range of length scales down to atomic dimensions. The development of chemically selective proximal probe imaging methods has played a central role in uncovering sample heterogeneity and understanding its origins. Numerous chemically selective, spectroscopic proximal probe methods continue to emerge from a number of labs around the world. Both the evolution of existing methods and the further development of new ones promise significant advances in our ability to obtain chemical information on heterogeneous samples on a variety of relevant length scales. One of the best examples of the use of proximal probe methods is the chemical bonding information that has been obtained on semiconductor surfaces by STM. However, in the biomedical realm and other application areas, the detection of discrete chemically specific binding interactions between proteins and peptides and/or drugs and receptors is of particular importance. Such measurements can be made by methods such as chemical force microscopy, but this approach lacks generality. That is, it provides significant specific information when much is already known about the sample surface composition. The development of new probes and chemical probe arrays will significantly advance these methods in the near future.

Contrast mechanisms that reveal chemical identity and function in surface characterization need to be improved for a wider variety of samples.

Near-field Optics

Molecular spectroscopies are usually restricted to length scales governed by the wave-like nature of light; specifically, spatial confinement of the source radiation is limited by the diffraction barrier to approximately one-half the wavelength of light. Near-field optical microscopy (optical proximal probe methods) overcomes this limitation and provides a means to extend optical spectroscopic techniques to the nanometer scale. However, the use of near-field microscopy to obtain chemical images of real-world samples remains hampered by issues of resolution and sensitivity. To overcome this limitation, we have to fabricate nanostructures that strongly localize and enhance the electric fields. Although recent results demonstrate the high potential of the field enhancement method, the technique is far from being well understood, reliable, or optimized.

There is a need to improve probe geometries for high-resolution chemical imaging beyond the diffraction limit. This includes design (theory) and realization (reproducibility, robustness, mass production) of controlled geometry near-field optics.

Image Processing and Analysis

Chemical imaging is used to selectively detect, analyze, and identify chemical and biological samples, followed by visualization of the data in the dimension of interest. The information of interest can range from composition, structure, and concentration to phase or conformational changes as a function of time or temperature. The expression "chemical image" describes a multidimensional dataset whose dimensions represent variables such as x, y, z spatial position, experimental wavelength, time, chemical species, and so forth. Image processing requires that the chemical images exist as digital images.

Initial Image Visualization

Frequently the first priority for the analyst is to generate an image or images that allow for visualization of heterogeneous chemical distributions in space or time. Image visualization methods vary from simply choosing a color scale for display of a single image to methods for displaying three-dimensional datasets. Simple gray-scale maps can be constructed from a single image. Different color scales can be utilized, and the contrast and brightness can be adjusted so that the information the analyst deems most important is emphasized. Multiple images from the same or different datasets can be viewed simultaneously for comparison. Scatter plots are frequently utilized for comparing two images. For more detailed comparisons among a small number of images, mapping individual images into red, green, and blue (RGB) channels creates composite color chemical images. For three-dimensional data, additional analysis tools are required, including the ability to extract spectra from a selected region of interest for multispectral imaging datasets or rendering a three-dimensional volume or projection for depth arrays.

Analysis tools for three-dimensional visualization need to be developed for various microscopies and materials analysis instrumentation.

Image Processing

Typically, data analysis is not considered until after an instrument is developed. This can often limit the imaging analysis or make it unnecessarily difficult. Particularly with quantitative techniques, questions such as, "How do you maintain calibration or correct for instrument drift over time?" have to be addressed from the initial stages of instrument design.

Researchers should be encouraged to integrate their data analysis with the development of their apparatus.

Multidimensional Image Processing

Present commercial multivariate analysis software is based on techniques that are more than 20 years old. Research on multivariate techniques, the development of chemometric analysis tools applied to imaging, and the deconvolution of hyperspectral images all need significant support. Developing these techniques to the point of routine use is an important challenge. Chemical imaging can benefit from other fields that depend on image analysis. The remote sensing field is rich in techniques for image analysis that are largely unexploited in chemical imaging; some kind of cross-fertilization between these two communities should be promoted. In addition, interactions with computer scientists and statisticians would enhance the development of chemical image analysis.

There is a need to develop better analysis and data extraction techniques for

elucidating more and different kinds of information from an image. In particular, this should include user-friendly multivariate analysis tools and hyperspectral imaging deconvolution and analysis.

Integrated Real-Time Analysis

As the sophistication and multimodality of imaging instrumentation increase and as the resolution of data collections improves, it will be necessary to perform some aspects of data reduction interactively during the measurement. Intelligent systems that can recognize information content and adjust resolution accordingly will have to be created. For example, high-resolution confocal imaging of large three-dimensional tissue sections in which specific, dispersed cell types are monitored for dynamic function requires programs that can automatically recognize appropriate cells and focus the dynamic data collection on those areas. **Integrated real-time analysis needs to be expanded for automated customization of data collection, particularly in multiscale imaging applications.**

Electronic Structure and Molecular Dynamics Simulations

A quantitative understanding of the electronic structure of molecules and the theory that predicts the outcome of interactions of molecules with electromagnetic fields will aid in the development of chemical imaging probes for all imaging modalities. For example, it has been known since the first NMR experiments that chemical shifts are exquisitely sensitive to the electronic environment of a molecule. The ability to understand electronic structure well enough to predict NMR chemical shifts should address a variety of problems in chemistry such as predicting reaction and folding pathways. **A quantitative understanding of molecular electronic structure is needed to make advances in chemical imaging. Two chief ways in which this understanding can be furthered are through improving probes and better theory.**

Molecular dynamics (MD) simulations can be performed to help better understand electronic and molecular structure. There exist at present several cybertools for performing MD simulations of various systems by experts or near experts. These tools are fairly mature in capability, but the user interfaces have only recently started to make them available to scientists other than the experts in the computational chemistry community.

Current MD simulations are thus not quite ready for use by nonexperts, particularly in cases involving multiple dimensions. One challenge will be to improve on these cybertools to make their use completely transparent. In particular, simulation packages taking MD approaches for modeling nonbiological systems lag behind comparable packages for biological systems. The packages should be able to interface with the common platforms (middleware) developed in the context of chemical imaging cyberinfrastructure. New theory, new algorithms, and new programs are needed to enable such use in the future. **There is a need to develop a next generation of readily accessible, easy-to-use MD simulation packages.**

In order to improve MD simulations, a number of specific areas should be addressed in the area of basic molecular dynamics theory. These include: (1) development of full quantum mechanical calculations on complex molecules and more robust ways to incorporate quantum mechanical calculations within larger-scale classical mechanics or statistical mechanics approaches; (2) development and refinement of transferable force fields between arbitrary atoms and molecules, which are necessary building blocks for MD simulations of general systems; and (3) development of multiscale theories and techniques for understanding systems. Moreover, the community must develop toolkits that allow general users to perform such simulations. **Chemical imaging would be invigorated by innovations in basic theory of molecular dynamics. At the same time, the specific needs of chemical imaging should play a role in guiding the development of MD theory.**

All Imaging Techniques

In addition to targeted improvements for individual and multitechnique approaches, there are certain developments needed for overall advances in chemical imaging capabilities. These range from improved light sources to data management, and they are discussed briefly below.

Light Sources

Advances in light sources are providing new capabilities in chemical imaging. For example, recent developments in free-electron lasers have led to a rapidly growing interest in using the terahertz range (3-300 cm^{-1}) for imaging. However, further development of terahertz spectroscopy as a powerful tool for imaging will depend on the development of convenient new terahertz light sources. **Brighter, tunable ultrafast light sources need to be developed, particularly infrared-terahertz vibrational and dynamic imaging, near-field scanning optical microscopy (NSOM), and X-ray imaging.**

Reverse Imaging

One of the emerging promises of chemical imaging is reversing the direction of information transfer. Many of the same approaches that are used to gather information at high resolution in spatial and temporal dimensions can also be used to control or manipulate chemical systems with similar spatial or temporal resolution. Examples include photolithography in the semiconductor industry and light-directed synthesis of oligonucleotides in the DNA chip industry. This could be greatly extended to include high-speed synthesis and activity screening, initiating or controlling chemical reactivity, spatially patterning cellular growth and function, and ultimately mediating chemical activity throughout an entire organism. **There is a need to develop imaging methods for patterning complex spatial and temporal organization into chemical systems.**

Optics

Lack of advances in optics has hampered improvements in microscopic imaging. Development of adaptable, inexpensive fiber optics to transmit high-energy femtosecond pulses from mode-locked lasers, custom phase plates, and miniature laser beam scanners for endoscopic microscopy instruments offer the potential for enormous advances in laser scanning microscopy for various applications, including medical diagnostics and surgery. **There is a need to develop optics for miniaturization and speeding of microscopic imaging instrumentation in order to improve chemical imaging capabilities.**

Acquisition Speed and Efficiency

Higher-speed scanning probes that now reach video-rate imaging have been developed. Brighter probes will reduce the integration time in single-molecule detection techniques. Increasing the speed of microscopic image application instrumentation offers the potential for rapidly advancing and expanding chemical imaging capability. **Acquisition speeds need to be increased in order to provide improved time resolution. In addition, there is a need to provide more online analysis capabilities to improve the efficiency of imaging by allowing more directed investigations of samples.**

Data Management

With increasing acquisition and imaging resolutions, data set size and volume, and processing capabilities, improved software to process and correlate images across experiments and time points will be needed. At the same time, cyber infrastructure and its underlying theoretical frameworks can provide computer-generated images that provide insight into the structure and function of both particular and generic samples. Simulations can also allow test cases for new paradigms for the cybertools used to manipulate data, process metrics, and render images using the experimental measurements. **Theory needs to be developed and better utilized to address the data storage**

and search problems associated with the increasingly large datasets generated by chemical imaging techniques.

INSTITUTIONAL CHANGE

As illustrated by the review of techniques and myriad examples given throughout this report, the field of chemical imaging is poised to provide fundamental breakthroughs in the basic understanding of molecular structure and function. The knowledge gained with these insights offers the potential for a paradigm shift in the ability to control and manipulate matter at its deepest levels. A strategic, focused research and development program in chemical imaging will best enable this transformation in our understanding of and power over the natural world. The committee has identified three key components of such a program: investigator approach, funding, and standards.

Investigator Approach

Some research efforts in chemical imaging (especially those at the interfaces of chemistry with other fields such as medicine, materials, and environment) may require a multidisciplinary team rather than individual investigator approach. For example, the realization of more safe and effective MRI contrast agents for clinical use requires a concerted effort of chemists, molecular biologists, radiologists, and MRI physicists. Forming a multidisciplinary team at an early stage and continued interaction at all stages of imaging technique development can be crucial. A key aspect of being able to form multidisciplinary teams is the development of reward systems at research institutions (funding, citations, and promotions or tenure) that recognize the equal importance of all facets of chemical imaging. **There is a need to encourage both individual investigator and multidisciplinary team approaches for the development of chemical imaging techniques.**

Funding

In many aspects, the resource appropriation for chemical imaging research works quite well. For example, support for synchrotron light sources is very effective and should be continued. As described in the previous paragraph, however, breakthroughs in imaging often require the sustained effort of multidisciplinary teams of scientists. This process, which is vital for success, does not naturally match present funding schemes and cycles. New programs and channels for funding have to be established to support this mode of science.

Many of the opportunities for imaging often require coordination across several funding institutions. For example, the Department of Energy holds the primary mission for the design, construction, and operation of major facilities such as light sources, while other agencies such as the National Science Foundation and National Institutes of Health support a significant number of science programs that utilize these facilities. Even greater cross-agency planning, coordination, and support will be vital to success. Support for this multidisciplinary research area in the form of (1) funding of proposals with multiple principal investigators, (2) financial support of the collaborative process (e.g. travel, meetings, training to facilitate interactions), and (3) larger grants of longer duration to finance extended startup periods could be crucial in facilitating chemical imaging development and advances.[3] In addition, creation of special funding channels directed exclusively toward chemical imaging development offers great promise for advances in this field.

Novel approaches to funding mechanisms for chemical imaging need to be promoted.

Standards

At present, each subfield of imaging science develops its own structure for data storage and archiving. This has resulted in countless formats that inhibit the sharing of data. A broad effort to develop standards for data format and organization across experimental contexts will give a great boost to imaging science. Standards are needed for chemical imaging that allows image data to be effectively catalogued and shared between large numbers of researchers. These data should be organized in visually oriented, searchable, and parameterized databases, with common platforms for a vast array of image types. The implementation of such a system will be transformational. The failure to implement such a system will mean that chemical image data obsolesce almost instantly upon their production. To avoid redundant effort and the squandering of resources, solutions must be found to ensure chemical image longevity.

There is a need to develop standards for chemical image data formatting.

CONCLUSION

Imaging has a wide variety of applications that have relevance to almost every facet of our daily lives. These applications range from medical diagnosis and treatment to the study and design of material properties in novel products. To continue receiving benefits from these technologies, sustained efforts are needed to facilitate understanding and manipulation of complex chemical structures and processes. Chemical imaging offers a means by which this can be accomplished by allowing the acquisition of direct, observable information about the nature of these chemistries. By linking technological advances in chemical imaging with a science-based approach to using these new capabilities, it is likely that fundamental breakthroughs in our understanding of basic chemical processes in biology, the environment, and human creations will be achieved.

NOTES

1. Kneipp K., H. Kneipp, P. Corio, S.D.M. Brown, K. Shafer, J. Motz, L.T. Perelman, E.B. Hanlon, A. Marucci, G. Dresselhaus, and M.S. Dresselhaus, 2000. Surface-enhanced and normal stokes and anti-stokes Raman spectroscopy of single-walled carbon nanotubes. *Phys. Rev. Lett.* 84: 3470-3473.

2. Hartschuh, A., E. J. Sanchez, X. S. Xie, and L. Novotny. 2003. High-resolution near-field Raman microscopy of single-walled carbon nanotubes. *Phys. Rev. Lett.* 90(9):095503.

3. National Research Council *Facilitating Interdisciplinary Research.* 2004. Washington, DC: The National Academies Press.

A

Statement of Task

The National Research Council, through its Board on Chemical Sciences and Technology, will review the current state of molecular imaging technology, point to promising future developments and their applications, and suggest a research agenda to enable breakthrough improvements in our capacity to image molecular processes simultaneously in multiple physical dimensions as well as time. This review will:

- Review the current state-of-the-art in chemical imaging, and identify likely short-term advances.
- Identify gaps in our knowledge of the basic science that enables chemical imaging. Discuss the advances that would be opened if these gaps were addressed.
- Utilizing this projection of future advances, develop a vision for the future of chemical imaging. Discuss the convergence of factors that make this vision timely. Identify major goals that could provide direction for prioritizing research aimed at advancing the field in the next 5 to 20 years.
- Identify research required to meet this vision. Include research required in probing molecular systems, detecting the resulting signals, and in manipulating and interpreting the resulting datasets, and discuss the instrumentation required to carry out this research. Consider both experimental and theoretical advances required. Focus particularly on areas where differential investment would have its greatest impact.
- Discuss educational and institutional innovations that could help catalyze advances in this field.

B

Committee Member Biographies

CHAIR

NANCY B. JACKSON is deputy director of the International Security Center at Sandia National Laboratories, where she is responsible for overseeing the international security programs, initiatives, and operations and coordinates the lab-wide initiative on Global Nuclear Futures, bringing together the nuclear energy and nonproliferation centers of the laboratory. Her technical experience lies primarily in the areas of spectroscopy and surface reactivity. Spectroscopic applications include fluorescence imaging of DNA microarrays, infrared imaging for analysis of polymeric materials, and point infrared spectroscopy for the study of structure-property relationships for heterogeneous catalytic materials. In addition, Dr. Jackson has managed research in the analysis of hyperspectral images for remote sensing and the hyperspectral analysis of materials using various chemical imaging instrumentation. Dr. Jackson earned her bachelor's degree in chemistry from George Washington University and her Ph.D. in chemical engineering from the University of Texas, Austin. Dr. Jackson is a member of the Board of Directors of the American Chemical Society and a fellow of the American Association for the Advancement of Science.

MEMBERS

PIERRE CHAURAND is currently research associate professor of biochemistry at Vanderbilt University (Nashville, TN). Among Dr. Chaurand's interests is research that combines cutting edge mass spectrometry technology and other technologies for profiling, identifying, and mapping the spatial distribution of bio-compounds directly in biological samples and the translation of these exciting new molecular technologies to the investigation of diseased tissues. In 1997, Dr. Chaurand received the annual young investigator prize from the French Mass Spectrometry Society. Dr. Chaurand obtained his Ph.D. in physical biochemistry and mass spectrometry in 1994 from the University of Paris Sud (Orsay, France).

JULIA E. FULGHUM currently serves as the chair of the University of New Mexico Chemical and Nuclear Engineering Department. Her research interests include materials characterization with an emphasis on multitechnique correlation and multivariate analysis for nondestructive evaluation of heterogeneous samples: X-ray photoelectron spectroscopy, time-of-flight secondary ion mass spectrometry, atomic force microscopy, Fourier-transform infrared spectroscopy. Prior to her current appointment, Dr. Fulghum was a faculty member in the Department of Chemistry at Kent State University (KSU) from August 1989 to August 2002, serving as an Honors College faculty member during her last two years. At KSU, she received the distinguished teacher award from the College of Arts and Sciences and was named outstanding faculty mentor in the Teaching Scholars Program, both in 2001. Dr. Fulghum serves as chair of the Advisory Board of the National ESCA (Electron Spectroscopy Chemical Analysis) and Surface Analysis Center for Biomedical Problems (NESAC/BIO) and is a member of the editorial advisory board for *Surface and Interface Analysis* and for the *Journal of Electron Spectroscopy and Related Phenomena*. In addition, she is active in the Applied Surface Science Division of the American Vacuum Society, where she has served as chair, program chair, and member-at-large. Dr. Fulghum received her Ph.D. in analytical chemistry in 1987 from the University of North Carolina, her master's degree in analytical chemistry in 1983 from Cornell University, and her bachelor's degree in chemistry with highest honors in 1981 from the University of North Carolina.

RIGOBERTO HERNANDEZ is associate professor of chemistry at Georgia Institute of Technology. His research interests focus on microscopic reaction dynamics and their effects on macroscopic chemical reaction rates in arbitrary solvent environments. His projects include the use of modeling to determine the chemical reaction dynamics of thermosetting polymers and living polymers, the diffusion of mesogens in a liquid crystal, the transport and control of adsorbates on a surface, the binding dynamics of proteins, and the dynamics of protein folding and rearrangement. Dr. Hernandez' awards include the Goizueta Foundation Junior Professorship (2002-2006); Sigma Xi Southeast Regional Young Investigator (2002); Alfred P. Sloan Fellow and Sigma Xi Southeast Regional Young Investigator (2000); Research Corporation Cottrell Scholar and Sigma Xi Young Faculty Award (1999); Blanchard Assistant Professorship of Chemistry (1999-2001); and National Science Foundation (NSF) CAREER Award (1997).

He was elected fellow of the American Academy of Arts and Science, in 2004. Dr. Hernandez received his B.S.E. from Princeton University in 1989 and Ph.D. from the University of California at Berkeley (1993). He was a Feinberg postdoctoral fellow at the Weizmann Institute of Science (1994) and a postdoctoral fellow at the University of Pennsylvania (1996).

DANIEL A. HIGGINS has served on the chemistry faculty at Kansas State University since 1996, where he currently holds the rank of associate professor. He conducts research involving the implementation of novel optical microscopic techniques for the characterization of mesostructured thin-film materials. He utilizes techniques such as near-field scanning optical microscopy (NSOM), single-molecule spectroscopy (SMS), and multiphoton-excited fluorescence microscopy to study organic polymer films, polymer-liquid crystal composites, polymer-surfactant complexes, and sol-gel derived silicate glass films. Dr. Higgins is a National Science Foundation (NSF) CAREER Award recipient and has received a 3M Company Untenured Faculty Research Award. Dr. Higgins received a B.A. from St. Olaf College (1988) and a Ph.D. from the University of Wisconsin, Madison (1993). He performed postdoctoral research at the University of Minnesota, where he held an NSF postdoctoral fellowship in chemistry.

ROBERT HWANG is the former director of the Center for Functional Nanomaterials at Brookhaven National Laboratory (BNL). He has recently returned to Sandia National Laboratories. Prior to his arrival at BNL, he managed the Thin Film and Interface Science Department at Sandia National Laboratories. He is currently a member of the Board on Chemical Sciences and Technology. Dr. Hwang received his B.S. from the University of California at Los Angeles and his Ph.D. from the University of Maryland.

KATRIN KNEIPP is currently an associate professor at Harvard Medical School, Wellman Center for Photomedicine, where she works on new spectroscopic methodologies for applications in life sciences and nanotechnology. Her current research focuses on combining modern laser spectroscopy with the exciting optical properties of nanostructures. By exploiting local optical fields of nanoparticles, molecular characterization and chemical imaging can be performed at the single-molecule level and from probed volumes on the nanoscale. Dr. Kneipp's research interests also include detection and structural characterization of single biomedically relevant molecules, as well as molecular probing in single living cells. Dr. Kneipp's work has been profiled in the *New Scientist* and the *American Institute of Physics Bulletin*. She received the Rockefeller-Mauze visiting chair award at Massachusetts Institute of Technology from 2000 to 2001 and the 1999 Meggers Award of the Society for Applied Spectroscopy. Dr. Kneipp received both her Ph.D. in physics and her Dr.Sc. in physical chemistry from the Friedrich-Schiller-University, Jena (Germany).

ALAN P. KORETSKY is chief of the Laboratory of Functional and Molecular Imaging and director of the National Institutes of Health (NIH) Magnetic Resonance Imaging (MRI) Research Facility at the National Institute of Neurological Disorders and Stroke (NINDS). Dr. Koretsky's laboratory is interested in two main areas: actively developing novel imaging techniques to visualize brain function and studying the regulation of cellular energy metabolism combining molecular genetics with noninvasive imaging tools. Dr. Koretsky spent 12 years on the faculty in the Department of Biological Sciences at Carnegie Mellon University, where he was the Eberly Professor of Structural Biology and Chemistry before coming to NINDS in 1999. Dr. Koretsky received his S.B. degree from the Massachusetts Institute of Technology and Ph.D. from the University of California at Berkeley. He performed postdoctoral work in the National Heart, Lung, and Blood Institute at NIH, studying regulation of mitochondrial metabolism using optical and nuclear magnetic resonance techniques. Dr. Koretsky was the recipient of the Gold Medal, the highest honor of the International Society of Magnetic Resonance, for his work in developing MRI tools to measure regional blood flow and image transgenic mice, and in the development of manganese-enhanced MRI.

CAROLYN LARABELL is professor of anatomy at the University of California, San Francisco, and concurrently holds a position in the Physical Biosciences Division at Lawrence Berkeley Laboratories. She conducts research using both electron and confocal microscopy and is currently developing the technology to obtain three-dimensional tomographic reconstructions of whole, hydrated cells at better than 50 nm isotropic resolution and to localize proteins and protein complexes in those cells using soft X-ray microscopy . Along with Dr. Mark Le Gros, she was awarded a $5.5 million grant from the Department of Energy and the National Institutes of Health to build a state-of-the-art X-ray microscope at Lawrence Berkeley National Laboratory to advance cellular and molecular biology and biomedical studies. Dr. Larabell received her Ph.D. in zoology from Arizona State University.

STEPHAN STRANICK is currently a senior scientist in the Chemical Science and Technology Laboratory of the National Institute of Standards and Technology. His present research focuses on the development of novel proximal probes of surface physiochemical properties, especially in chemical imaging, using near-field optical microscopy. He has published more than 50 papers on the subject and has been awarded 14 patents associated with scanned probe microscopies for chemical and electrical characterization. Dr. Stranick's awards include the American Chemical Society's Nobel Laureate Signature Award, the American Chemical Society's Procter & Gamble Award in Physical Chemistry, the American Chemical Society's Arthur F. Findeis Award for Achievement by a Young Analytical Scientist, a BF Goodrich Inventors Award, the Union Carbide Kenan Analytical Award, the Xerox President's Award in Materials Research, the

Samuel Wesley Stratton Award for Excellence in Science and Engineering, and the Department of Commerce Bronze Metal Award for Superior Federal Service. Dr. Stranick received his B.S. in chemistry from Ithaca College (1989) and his Ph.D. in chemistry from Pennsylvania State University (1995).

WATT WEBB joined the Cornell faculty in 1961, served as director of the School of Applied and Engineering Physics from 1983 to 1989, and is currently a member of the graduate faculties of seven fields. In addition to directing the Developmental Resource for Biophysical Imaging Opto-electronics, he serves on the board of directors and executive committee of the Cornell Research Foundation. Dr. Webb is also affiliated with the university's Biophysics Program, the Cornell Center for Materials Research, and the National Biotechnology Center and serves on the Life Sciences Advisory Council. He is a fellow of the American Physical Society (APS) and the American Association for the Advancement of Science; a founding fellow of the American Institute of Medical and Biological Engineers; and an elected member of the National Academy of Engineering, National Academy of Sciences, and American Academy of Arts and Sciences. He won the APS Biological Physics Prize in 1990, the Ernst Abby Lecture Award in 1997, the Michelson-Morley Award in 1999, and the Rank Prize for Opto-electronics in 2000; he was the Jablonski Award lecturer in 2001 and the 2002 national lecturer of the Biophysical Society, and he has served as chairman of the Division of Biological Physics and associate editor of *Physical Review Letters*. Currently, his studies involve new experimental technologies to study the dynamics of biomolecular life processes, including multiphoton microscopy, fluorescence correlation spectroscopy, and nanoscopic molecular tracking. Dr. Webb received both his B.S. (1947) and his Sc.D. (1955) from the Massachusetts Institutes of Technology.

PAUL WEISS currently serves as distinguished professor of chemistry and physics at the Pennsylvania State University, where he began his academic career in 1989. His research efforts are focused on gaining atomic-scale understanding and control of materials properties, and his group performs exploration, probing, and manipulation of the interactions and dynamics at surfaces and interfaces using scanning tunneling microscopy and related techniques. He advances scanning probe microscopy and nanolithography through developing new methods and capabilities in each. He and his group also develop and explore physical models of biological systems. Dr. Weiss received both his bachelor's and master's degrees in chemistry from the Massachusetts Institute of Technology (1980) and his Ph.D. in chemistry from the University of California at Berkeley (1986). He was a postdoctoral member of the technical staff at AT&T Bell Laboratories (1986-1988) and a visiting scientist at IBM (Almaden Research Center, 1988-1989). Dr. Weiss' awards and honors include the Scanning Microscopy International Presidential Scholarship (1994), the B.F. Goodrich Collegiate Inventors Award (1994), the

American Chemical Society Nobel Laureate Signature Award for Graduate Education in Chemistry (1996), and the National Science Foundation Creativity Award (1997-1999); he was elected a fellow of the American Association for the Advancement of Science (2000) and the American Physical Society (2002).

NEAL WOODBURY joined the faculty of Arizona State University in 1987, where he is currently professor of chemistry and biochemistry. In addition, he serves as director of the university's Center for BioOptical Nanotechnology in the Biodesign Institute and served as the director of the Photosynthesis Center from 1997 until 2000. Dr. Woodbury was a pioneer in using photophysical principles to study the mechanism and dynamics of biochemical reactions such as photosynthetic energy conversion. By using laser technology and focusing on reactions initiated by light, Dr. Woodbury and his team made significant progress in understanding ways in which energy can be harnessed from light to both probe and manipulate biological reactions. An advocate of interdisciplinary science, he believes that a broad-based understanding of biology, chemistry, and physics provides researchers greater vision in addressing real-world problems. To this end, he directs a National Science Foundation Integrative Graduate Education and Research Traineeship (IGERT) program, bringing together students from physical sciences, life sciences, and engineering. He is an active member of the American Chemical Society, Biophysical Society, and American Photobiology Society and is coauthor of more than 75 published articles and studies. Dr. Woodbury received his B.S in biochemistry from the University of California, Davis (1979), and his Ph.D. in biochemistry from the University of Washington (1986).

XIAOLIANG SUNNEY XIE is professor of chemistry and chemical biology at Harvard University, where his research is focused on imaging, spectroscopy, and the dynamics of single biomolecules and single cells. He and his research group did pioneering work on fluorescence studies of single molecules at room temperature, near-field microscopy, single-molecule enzymology, and coherent anti-Stokes Raman scattering microscopy. Prior to joining the faculty of Harvard, Dr. Xie was a chief scientist in the Environmental Molecular Sciences Laboratory at Pacific Northwest National Laboratory. Dr. Xie received his B.S. in chemistry from Peking University (1984) and his Ph.D. in chemistry from the University of California at San Diego (1990); he did his postdoctoral work at the University of Chicago (1990-1991). He currently serves on the advisory boards of several government and academic institutions, as well as several journals including *Annual Review of Physical Chemistry*, *Accounts of Chemical Research*, and *Journal of Physical Chemistry*. He has coauthored more than 100 papers and holds three patents. Past awards include the National Institute of Health's Director's Pioneer Award (2004), the Raymond and Beverly Sackler Prize in the Physical Sciences (2003), and the Coblentz Award from the Coblentz Society (1996).

EDWARD YEUNG is on the chemistry faculty at Iowa State University, where he is currently distinguished professor in liberal arts and sciences. His research interests span both spectroscopy and chromatography, with publications in areas such as nonlinear spectroscopy, laser-based detectors for liquid chromatography, capillary electrophoresis, trace-gas monitoring, single-cell and single-molecule analysis, DNA sequencing, and data treatment procedures in chemical measurements. He is an associate editor of *Analytical Chemistry* and has served on the editorial advisory board of *Progress in Analytical Spectroscopy, Journal of Capillary Electrophoresis, Microchimica Acta, Spectrochimica Acta Part A, Journal of Microcolumn Separations, Electrophoresis, Journal of High Resolution Chromatography, Chromatographia,* and *Journal of Biochemical and Biophysical Methods.* Past awards include the Alfred P. Sloan Fellowship in 1974, the American Chemical Society Awards in Analytical Chemistry (1994) and Chromatography (2002), the International Prize of the Belgian Society of Pharmaceutical Sciences (2002), and election as a fellow of the American Association for the Advancement of Science in 1992. Dr. Yeung received his A.B. degree in chemistry from Cornell University in 1968 and his Ph.D. in chemistry from the University of California at Berkeley in 1972.

C

Guest Panelists

HAROLD W. ADE, North Carolina State University
ASHOK A. DENIZ, The Scripps Research Institute
KENNETH DOWNING, Lawrence Berkeley Laboratories
NIGEL GOLDENFELD, University of Illinois at Urbana-Champagne
TAEKJIP HA, University of Illinois at Urbana-Champagne
DAVID HAALAND, Sandia National Laboratories
WILSON HO, University of California, Irvine
ALAN HURD, Los Alamos National Laboratories
CHRIS JACOBSEN, State University of New York at Stony Brook
SUDHIR KUMAR, Arizona State University
STUART LINDSAY, Arizona State University
LUKAS NOVOTNY, University of Rochester
DANIEL RUGAR, IBM Almaden Research Center
JONATHAN V. SWEEDLER, University of Illinois at Urbana-Champagne
DEVARAJAN THIRUMALAI, University of Maryland
RUDOLF TROMP, IBM T.J. Watson Research Center
ROGER TSIEN, University of California, San Diego
RICHARD P. VAN DUYNE, Northwestern University

D

Acronyms and Abbreviations

ACRONYMS

ADMP	Atom-centered Density Matrix Propagation
AES	Atomic Emission Spectroscopy
AFM	Atomic Force Microscopy
ATP	Adenosine 5'-Triphosphate
AXSIA	Expert Spectral Image Analysis
BEEM	Ballistic Electron Emission Microscopy
BOMD	Bohr-Oppenheimer Molecular Dynamics
CARS	Coherent Anti-Stokes Raman Scattering
CAT	Computerized Axial Tomography
CCD	Charge-Coupled Device
CIF	Chemical Image Fusion
CPMD	Carr-Parrinello Molecular Dynamics
CT	Computed Tomography
CTAB	Cetyltrimethylammonium Bromide
CW	Continuous Wave (Laser)
DCT	Discrete Cosine Transformation
DIC	Differential Interference Contrast
DICOM	Digital Imaging Communications in Medicine
DIM	Diffraction Imaging Microscopy
DWT	Discrete Wavelet Transform

EDXS	Energy-dispersive X-ray Spectroscopy
EELS	Electron Energy Loss Spectroscopy
EM	Electron Microscopy
ENVI	Environment for Visualizing Images
FCS	Fluorescence Correlation Spectroscopy
FT	Fourier Transform
fMRI	Functional Magnetic Resonance Imaging
FTIR	Fourier Transform Infrared Spectroscopy
GFP	Green Fluorescent Protein
GID	Global Image Database
GISAXS	Grazing Incidence Small-angle X-ray Scattering
GMR	Giant Magnetoresistivity
GRIN	Gradient Refractive Index
GUI	Graphical User Interface
HPLC	High-performance Liquid Chromatography
ICS	Indosyanine Green
IETS	Inelastic Electron Tunneling Spectroscopy
IXS	Inelastic X-ray Scattering
IR	Infrared
LCLS	Linear Coherent Light Source
LDL	Low-density Lipoprotein
LED	Light-emitting Diode
LEEM	Low-energy Electron Microscopy
MALDI	Matrix-assisted Laser Desorption/Ionization
MCM-41	Mobile Crystalline Material-41
MCR	Multivariate Curve Resolution
MD	Molecular Dynamics
ME	Matrix-enhanced
MFM	Magnetic Force Micrscopy
MI	Multivariate Image
MPM	Multiphoton Microscopy
MRI	Magnetic Resonance Imaging
MS	Mass Spectroscopy
MVA	Multivariate Analysis
NEXAFS	Near-edge X-ray Absorption Fine Structure (Spectroscopy)
NIBIB	National Institute of Biomedical Imaging and Bioengineering

NIGMS	National Institute of General Medical Sciences
NIH	National Institutes of Health
NIR	Near Infrared
NMR	Nuclear Magnetic Resonance
NRS	Nuclear Resonant Scattering
NSOM	Near-field Scanning Optical Microscopy
PACS	Picture Archiving and Communication Systems
PCA	Principal Component Analysis
PDB	Protein Data Bank
PEEM	Photoemission Electron Microscopy
PESTM	STM Photoemission Spectroscopy
PET	Positron Emission Tomography
PrP	Prion Protein
RFP	Red Fluorescent Protein
RGB	Red, Green, Blue
ROI	Region of Interest
RIXS	Resonant Inelastic X-ray Scattering
SAXS	Small-angle X-ray Scattering
SEM	Scanning Electron Microscopy
SERS	Surface-enhanced Raman Scattering
SHG	Second Harmonic Generation
SIMS	Secondary Ion Mass Spectrometry
SPM	Scanning Probe Microscopy
SQUID	Superconducting Quantum Interference Device
STEM	Scanning Transmission Electron Microscopy
STM	Scanning Tunneling Microscopy
STXM	Scanning Transmission X-ray Microscopy
SXES	Soft X-ray Emission Spectroscopy
TEM	Transmission Electron Microscope
TERS	Tip-enhanced Ramen Spectroscopy
THG	Third Harmonic Generation
THz	Terahertz
TOF	Time of Flight
TXM	Full-field X-ray Microscopy
UV	Ultraviolet
WAXS	Wide-angle X-ray Scattering
WT	Wavelet Transform